The Earth's Electric Field

The Earth's Electric Field

Sources from Sun to Mud

MICHAEL C. KELLEY
School of Electrical and Computer Engineering
Cornell University
Ithaca, New York

with contributions from

ROBERT H. HOLZWORTH
Earth and Space Sciences
University of Washington
Seattle, Washington

ELSEVIER

AMSTERDAM • BOSTON • HEIDELBERG • LONDON • NEW YORK • OXFORD
PARIS • SAN DIEGO • SAN FRANCISCO • SINGAPORE • SYDNEY • TOKYO

Elsevier
225 Wyman Street, Waltham, MA 02451, USA
525 B Street, Suite 1900, San Diego, CA 92101-4495, USA

First edition 2014

Library of Congress Cataloging-in-Publication Data
A catalog record for this book is available from the Library of Congress
Kelley, Michael C.
  The earth's electric field: sources from sun to mud / Michael C. Kelley. – First edition.
    pages cm
  Includes bibliographical references.
  ISBN 978-0-12-397886-8
  1. Terrestrial radiation. 2. Electromagnetic fields. 3. Electromagnetic waves. 4. Atmospheric
electricity. I. Title.
  QC809.T4K45 2013
  538′.72–dc23

                                                                                    2012049174

British Library Cataloguing in Publication Data
A catalogue record for this book is available from the British Library

For information on all **Elsevier** publications
visit our web site at store.elsevier.com

Printed and bound in Sheridan Books
14 15 16 17 18   10 9 8 7 6 5 4 3 2 1

ISBN: 978-0-12-397886-8

Dedication

To my beloved family

Aidan, Amelia, Brian, Elizabeth, Erica, Owen, Patricia, Scott, and Varykina.

CONTENTS

Knowledge of the earth's electric field has grown greatly during the past few decades. Prior to 1960, the primary interest was in the atmospheric electric field, the fields generated in thunderclouds, and the currents with which they charge the earth worldwide. These studies are most famously linked to Ben Franklin. In the 1960s, the development of scientific radars, scientific rockets, and satellites extended our knowledge into space. Rockets, radars, and low-altitude satellites probed the ionosphere, roughly 100-1000 km in altitude. Satellites extended our knowledge into the magnetosphere, a vast region dominated by the earth's magnetic field and then into the region called the solar wind, which is dominated by the sun's upper atmosphere. We treat each of these various areas in this text.

The history of magnetic field studies is much, much longer. For millennia, humankind has known of and used the earth's magnetic field. Long ago, Chinese and later, European explorers used the earth's poles as guides into uncharted waters. Today, scientists have developed far more complex models of the earth's magnetic field and continuously monitor magnetic fluctuations to push our understanding still further. The earth's magnetic field has been likened to one produced by a bar magnet lying at an 11° angle with the spin axis of the earth. However, while the field lines close to the earth closely resemble those of a bar magnet, field lines are greatly distorted by solar winds at greater distances from the earth's surface.

While the source of a bar magnet's magnetic field is aligned electron magnetic moments, each contributing to the total magnetic field, this cannot be the cause of the earth's magnetic field. The temperature in the earth's core is simply too hot, well above the Curie point for iron. Above this temperature, atoms have so much energy that electron spins no longer align and are more or less random, resulting in a net magnetic field of zero. Scientists believe that telluric currents below the surface of the earth are the source of the earth's magnetic field. The source of telluric currents has been explained by the dynamo effect. For quite some time, scientists have accepted the dynamo effect as the source of the earth's magnetic field but now are using complex computer models to provide conclusive proof. Only now are we moving closer to exploring the electric field in models of this sort.

Many books have been written about the earth's magnetic field. In fact, *Epistola de Magnete* (1269) by Peregrinus is considered to be the first scientific paper ever written. An additional important text, *De Magnete* (1600), was written by William Gilbert.

In this text we try, for perhaps the first time, to offer a systematic approach to describing the earth's electric field. In Chapter 1, we discuss how electric fields are generated using tools common to junior-level electrophysics concepts. Chapter 2 examines the atmospheric electric field sources that dominate below about 90 km. In the 90-1000 km height range, what we term a hydromagnetic generator operates, as discussed in Chapter 3. In this case, the source of energy is the high-altitude atmospheric wind which, when it blows across the magnetic field, generates electric currents. The current system is very complex, charges must build up on boundaries, and electric fields result. Because of the high conductivity along magnetic field lines, which act like electric wires, electric fields are mapped for vast distances, even hemisphere to hemisphere. This generator dominates below about 60° latitude.

In Chapter 4, we turn to fields generated by the sun's upper atmosphere, which expands at supersonic speeds past the earth, sweeping the earth's magnetic field into a huge comet-like shape called the magnetosphere. This solar wind is a highly conducting plasma and hence any electric field is shorted out. However, in the earth's frame of reference, there is quite a large field, called a hydromagnetic generator, that generates up to a few million volts across the magnetosphere. About half of the time, depending on the interplanetary magnetic field direction, roughly 10% of this voltage penetrates into the earth's polar regions and drives the magnificent aurora, huge electrical currents, and many fascinating atmospheric phenomena. This penetration occurs on the earth's magnetic field lines that connect to the solar magnetic field above about 75° latitude. Between 60° and 75°, the magnetic field lines connect between the hemispheres but are stretched out of the dipole shape. In Chapter 5, we describe the physical process of a secondary hydromagnetic generator in this latitude region. Again, many interesting phenomena occur due to the currents, electric fields, and plasma boundary phenomena. In Chapter 6, we introduce electromagnetic and electrostatic wave phenomena, which are ubiquitous in the earth's environs. In Chapter 7, we review the most important electric field measurement techniques.

Hopefully, this unified description will be a useful introduction to the fascinating electrical processes around the earth.

Michael C. Kelley and Robert H. Holzworth

Electric Field Generation Mechanisms

The Earth's Electric Field. http://dx.doi.org/10.1016/B978-0-12-397886-8.00001-6   1

1.1 INTRODUCTION

We live in an electromagnetic world, from the chemical bonds that hold us together to the large-scale, powerful effects of thunderstorms and solar flares. Human understanding of electricity began during a time of fascination with and fear of unexplained natural phenomena such as the aurora borealis and lightning. Hundreds of years ago, the earth's magnetic field first became a useful tool for explorers, yet even today, there is no similarly useful large-scale map of the earth's electric field. Indeed, while the earth's magnetic field lines are continuous and penetrate even the depths of the earth, electric fields begin and end with electric charges, which can be anywhere.

Lacking an accurate global electric field model to complement the one we do have for the earth's magnetic field, we still can gain critical insight into our electrical environment by applying basic electromagnetic theory to our geophysical environment. At the most fundamental level, charged particles, such as electrons and ions, cannot gain energy unless an electric field exists in their frame of reference. Acceleration in circular motion, such as an orbiting satellite or an electron encountering a static magnetic field, does not result in energization of the satellite or the electron. Rather, circular motion causes only a continuous change in the object's direction of motion, not the magnitude of the velocity. Thus, imparting energy to a charged particle fundamentally requires electric fields and forms the most basic reason for studying geophysical electric fields.

A major complicating factor in studying the earth's electrical environment is that strong electric fields can exist in a region with practically no evidence of these strong fields outside that localized region. For instance, we now know that large-scale electric fields dominate the dynamics of near space, including the ionosphere and magnetosphere. Yet, these electric fields are nearly impossible to detect near the ground, just 100 km from the global-scale sources in the ionosphere. It was not until the 1960s, shortly after Sputnik and the beginning of the Space Age, that we began to learn anything about these electric fields. Compared to magnetic field research, environmental electric field research in our near-space region is hundreds of years behind. Of course, back in the 1750s, Ben Franklin first identified the large electric fields associated with thunderstorms and lightning. Within decades, this new knowledge changed our world view (c.f. Krider), removing a layer of superstition about the nature of lightning. Even today, the electrical potentials found inside thunderstorms, which can exceed a billion volts (1,000,000,000 V) across a few kilometers (Tom Marshall article), remain hard to predict and model accurately because measurements inside thunderstorms are extremely difficult to make.

Knowledge of electric fields in the atmosphere and space has become very important to our understanding of the dynamics of both the neutral and charged atmosphere at all altitudes, from the clouds to outer space. Even short distances away from the huge fields in thunderstorms, the surface electric field may bear little or no resemblance to fields in the charged cloud. Thus, we depend on *in situ* measurements of the vector electric field by balloon, rocket, and satellite platforms to enhance our understanding of the earth's electric field.

1.2 VISUAL EFFECTS OF ELECTRICAL PHENOMENA

When the electrons of an atom or molecule relax into lower energy states, they emit electromagnetic energy in the form of photons. These photons often have wavelengths in the visible spectrum, allowing us to "see" the electrical processes in action. Similarly, when an ion recombines with an electron, it also emits a photon, allowing charged gas dynamics to become visible. A familiar example is the little spark one receives when touching a doorknob on a low-humidity day. The process of spark discharge occurs when the electric field between your finger and the doorknob exceeds 30,000 V/cm! This is the electric field required to break down the air, allowing a cascade of electrons

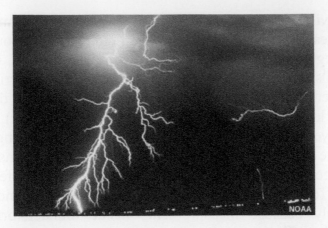

Fig. 1.1 A nighttime photograph of multiple cloud-to-ground and cloud-to-cloud lightning strokes. Thousands of flashes, each consisting of several strikes, occur every day around the world. (For a color version of this figure, the reader is referred to the online version of this chapter.) Figure courtesy of NOAA.

to flow and neutralize the electric charges that caused the large electric field in the first place. When energetic electrons pass through neutral air, they can cause ionization (as for the spark) or simply excitation of the air molecules. The visible aurora borealis is an example of excited molecules or atoms energized by the impact of magnetospheric electrons and ions, which subsequently release that energy as visible photons (Fig. 1.1).

In this section, we examine visible evidence of the earth's large-scale electric fields to set the stage and motivate the technical study found in this book. Lightning is primarily a land-based phenomenon. Each cloud-to-ground (CG) flash actually consists of several strokes that follow the same channel. However, each stroke lasts less than a millisecond and the whole flash typically less than a second. This time frame is usually too small for the human eye to clearly resolve each flash into multiple events. A typical multi-stroke flash lasts several hundred milliseconds but can last longer than 1 s. The majority of CG lightning strokes bring negative charge to the earth, leaving net positive charge behind in the cloud. A few percent of all CG strokes have the opposite polarity, depositing positive charge on the earth. (For more information about the physics of lightning, see Rakov and Uman, 2003.)

Recently, it was discovered that the electric fields inside clouds can be large enough to accelerate particles so high that, when they collide with atmospheric particles, they not only ionize but also create antimatter (positrons). Using data from the NASA Fermi satellite, a team under Briggs studying high-energy photon bursts associated with lightning, called terrestrial gamma-ray flashes (TGFs), revealed the characteristic decay signature of positrons during high-altitude lightning events (Briggs et al., 2010, 2011). Using lightning data from the World Wide Lightning Location Network, the Fermi satellite group demonstrated that these TGFs are associated with in-cloud discharges, producing extremely high currents composed of both electrons and positrons streaming oppositely to make electric currents that are orders of magnitude larger than ever thought possible with just electrons (Connaughton et al., 2013) (Fig. 1.2).

Above the cloud, we find that positive lightning strokes can also create another spectacular light show in the upper atmosphere, referred to as TLEs for transient luminous events. One type of TLE is called a sprite, as shown in Fig. 1.3. Sprites are very bright and equally short-lived, lasting only a few milliseconds. These events initially develop visible light near 70 km altitude where breakdown first starts, with subsequent optically visible extensions upward and downward from that altitude. Although some pilot reports of TLEs had occurred, it was not until 1989 that the first sprite was imaged by Winckler's group (see Franz et al., 1990). Following this first actual recorded image of a sprite, space shuttle video images, taken by a hand-held imager, were examined for sprites, and several were found by Vaughan's group (Boeck et al., 1992). However, the scientific community did not start concerted study of the phenomena related to TLEs until Lyons recorded multiple sprites in 1992, imaged from his home while looking out over the Great Plains (see Lyons et al., 1994). After the

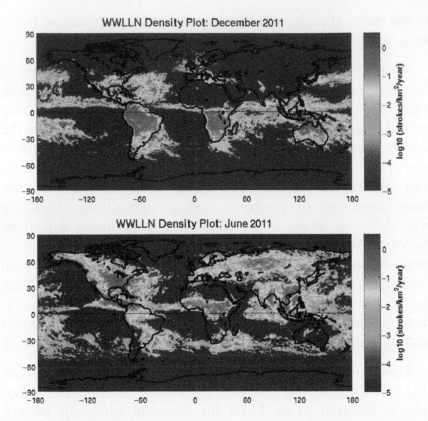

Fig. 1.2 World Wide Lightning Location Network (WWLLN) lightning density plots for June and December 2001 showing the sharp contrast in the Northern and Southern Hemispheric lightning activity with seasons. In December, continental lightning below the equator in South America, Africa, and Indonesia dominates the global lightning activity, whereas in Northern Hemispheric summer, we have copious lightning all the way up through Alaska and Siberia. For more information, see the WWLLN Web sites. (For a color version of this figure, the reader is referred to the online version of this chapter.) Images reproduced with permission from the WWLLN and courtesy of M. L. Hutchins, University of Washington.

community had a reliable method for imaging sprites, the field of TLE research really took off with the subsequent discovery of many new related phenomena called blue jets, elves, halos, giant jets, etc. A summary of some of the TLE lightning effects in the upper atmosphere is presented in Fig. 1.4.

A beautiful but far less destructive phenomenon, the aurora, is caused by the interaction of highly energetic particles and the earth's atmosphere. Charged particles from the solar wind enter

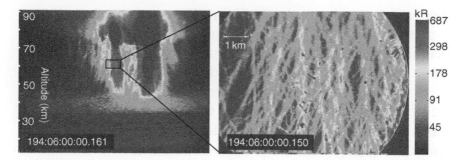

Fig. 1.3 Wide (left-hand panel) and telescopic (right-hand panel) field-of-view (FOV) false-colored images of a bright sprite event over Mexico. The narrow FOV with respect to the wide FOV is outlined by a black rectangle in the left-hand panel. This telescopic image demonstrates that, while the sprite appears to be amorphous in the wide FOV image, the lower portion of this sprite actually consists of a volume of densely packed filaments. (For a color version of this figure, the reader is referred to the online version of this chapter.) Adapted from Gerken et al. (2000). Reproduced with permission of the American Geophysical Union.

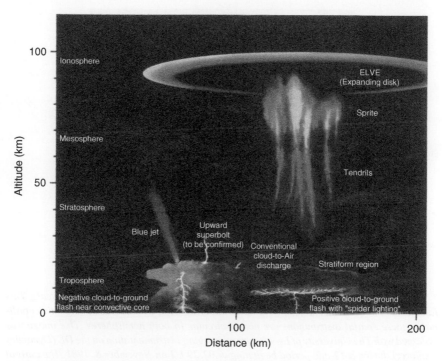

Fig. 1.4 Schematic of upper atmospheric transient luminous events (TLEs) related to lightning strokes. (For a color version of this figure, the reader is referred to the online version of this chapter.)

the earth's magnetosphere. Through a series of cascading events, particles in the earth's magnetosphere are forced by magnetic pressure and tension toward the earth. Once within about six earth radii (R_e) of the surface (1 $R_e = 6390$ km), the electrons spiraling along the magnetic field lines are accelerated by an electric field parallel to the magnetic field. They reach about 5000 V and impact the atmosphere where light is emitted in a magnificent show known as the aurora. The aurora courses virtually continuously over the northern and southern high-latitude zones. This continuously glowing spectacle is shown in a series of photos of the auroral oval in Fig. 1.5. The expansion in size of the oval seen in the figure mimics the increase in energy input under certain interplanetary conditions. The development of structure in the oval occurs when stored energy in the system is unloaded into the atmosphere.

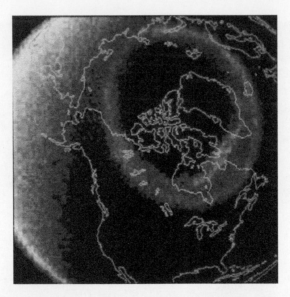

Fig. 1.5 The spatial distribution of instantaneous auroral emissions on a global scale. This false-color image shows vacuum-ultraviolet wavelengths of the aurora. In magnetic coordinates, these spatial distributions are nearly circular in both hemispheres. This image was obtained with The University of Iowa's auroral imaging instrumentation on the DE (Dynamics Explorer) during a 12-min period beginning at 02:29 UT on November 8, 1981. The auroral oval approaches the terminator at local noon. For good videos of the aurora, see The Gateway to Astronaut Photography of Earth on the NASA.gov site. (After Craven and Frank, 1991) (For a color version of this figure, the reader is referred to the online version of this chapter.)

The size of the auroral oval is a direct measure of the energy input from the solar wind. In the earth's frame of reference, the solar wind presents an electric field that is driven by what we term a collisionless magnetohydrodynamic (MHD) generator, as described in Chapter 4. A secondary, collisionless MHD generator is located inside the earth's magnetosphere and is studied in Chapter 5. A view of the aurora from 800 km, along with the city lights in the Eastern United States, is given in Fig. 1.6. Hudson Bay is seen at the top of the image.

Motions in the earth's upper atmosphere act in a way that is termed a collisional hydrodynamic generator. This source dominates the electric field below about 60° magnetic latitude and is described in Chapter 3. The solar wind, magnetospheric, and atmospheric generators communicate with the earth's ionosphere and atmosphere, creating a coupled atmosphere-ionosphere-magnetosphere dynamic system. Wave phenomena are often important sources of electric fields in space and in the atmosphere and are discussed in Chapter 6. Electric field measurement techniques are discussed in Chapter 7.

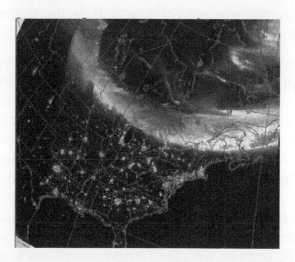

Fig. 1.6 The bright arc stretching from Montana to Maine is the aurora borealis. In many places, it completely overwhelms the city lights below. This image is a mosaic of various DMSP overflights on December 12-13, 13-14, and 14-15, 2006. (For a color version of this figure, the reader is referred to the online version of this chapter.) Figure courtesy of SpaceWeather.com.

1.3 GENERATION OF ELECTRIC FIELDS

The author assumes only a basic knowledge of electrodynamics and fluid dynamics at the sophomore college physics level. Some basics are reviewed here and then specialized equations are derived as needed "just in time" in the text. A more complete derivation of almost every equation used here can be found in Kelley (1989, 2009).

How are electric fields generated in any system? The deceptively simple Maxwell equations give the answer. Electric charges and time changes in the magnetic field are the two ways in which electric fields are generated, as shown in the following two equations from the Maxwellian set:

$$\nabla \cdot \mathbf{E} = \frac{\rho}{\varepsilon_0}, \tag{1.1}$$

$$\nabla \times \mathbf{E} = -\frac{\partial \mathbf{B}}{\partial t}, \tag{1.2}$$

where $\mathbf{E}$ is the electric field (volts/meter $=$ V/m), ρ is the charge density (Coulombs/m$^3 =$ C/m^3), and $\mathbf{B}$ is the magnetic field (Tesla $=$ T), using MKS units.

Unfortunately, Eq. (1.1) is not very useful. The problem is that $\rho = (Z_i n_i - n_e)e$, the charge density difference between ions with a net number of charges Z_i (usually $Z_i = 1$) and electrons, can be exceedingly small and yet generate substantial electric fields. In this section, we will assume $Z_i = 1$ until we come to the lowest part of the atmosphere. For example, a large ionospheric electric field is 0.1 V/m. This can be estimated and generated by a charge density of about 10^{-12} C/m^3. This is about one part in 10^8 for a typical ionospheric daytime number density. Such a large disparity is impossible to deal with experimentally or theoretically, or even with the largest computers available.

Since Eq. (1.1) is not very helpful, we are in big trouble. But we also have the principle that charge is conserved. To use this fact, consider a closed volume. If a current flows across its surface into the volume, the charge must increase inside. A simple but useful analogy is the flow of water. If there is a net flow of water into a closed surface, the amount inside clearly must increase. In mathematical terms, using integral calculus,

$$\iint_{\Sigma} \mathbf{J} \cdot d\mathbf{s} = -\iiint \frac{\partial \rho}{\partial t} dV,$$

where $\mathbf{J}$ is the current density with units Amperes/m^2 (C/m^2 s). The surface integral of the current density projected across the boundary at each surface area equals the volume integral of the charge density increase inside. Comparing the units,

$$\frac{C}{sm^2}\left(m^2\right) = \frac{C}{sm^3}\left(m^3\right),$$

which checks. The minus sign above is due to d$\mathbf{s}$ being pointed out of the volume. The integral form for equations such as this is intuitive and hence relatively easy to understand. But integral equations are very difficult to solve and must be converted to differential forms. This is accomplished using the Gauss and Stokes theorems. Here, we use Gauss's theorem,

$$\iint_{\Sigma} \mathbf{F} \cdot d\mathbf{s} = \iiint (\nabla \cdot \mathbf{F}) dV$$

where, in Cartesian coordinates,

$$\nabla \cdot \mathbf{F} = \frac{\partial F_x}{\partial x} + \frac{\partial F_y}{\partial y} + \frac{\partial F_z}{\partial z}.$$

For reference, Stokes' theorem is

$$\iint_{\Sigma} (\nabla \cdot \mathbf{F}) \cdot d\mathbf{s} = \int \mathbf{F} \cdot d\mathbf{r}$$

where the latter is a line integral around the boundary of the surface Σ. Applying Gauss' theorem to the double integral of $\mathbf{J}$ above,

$$\iint_{\Sigma} \mathbf{J} \cdot \mathbf{ds} = \iiint (\nabla \cdot \mathbf{J}) \mathrm{d}V.$$

But this means that

$$\nabla \cdot \mathbf{J} = \frac{-\partial \rho}{\partial t}. \tag{1.3}$$

This is also a statement of the conservation of charge. That is, to change the charge density in a volume, charge must flow into or out of that volume. Even generating new charge inside the volume (for instance, by ionization) does not change the total charge density ρ because new, free charge is generated in positive and negative pairs. In the author's opinion, physical interpretation of the differential form is more difficult than for the integral form. It is interesting to note that to develop phenomena of this sort experimentally, laboratory conditions that are more closely described by the integral mathematical form are utilized. For example, consider the electric field of a point charge. Experiments in the 1800s showed that the field is proportional to the charge of g, decreases with distance as $1/r^2$, and is radially outward,

$$E(r) = \frac{Kq}{r^2}.$$

Thus, the experimental data allow easy evaluation of the surface interval in Gauss' theorem and leads to the differential form of one of Maxwell's equations,

$$\nabla \cdot \mathbf{E} = \frac{\rho}{\varepsilon_0}.$$

Similarly, the magnetic field of a long wire carrying current I is found to be proportional for I/r in the azimuthal direction.

Thus, the integral can be evaluated experimentally. Then Stokes' theorem leads to another of Maxell's equations:

$$\nabla \cdot \mathbf{B} = 0.$$

Thus, the experimental integral approach leads to the more mathematically useful differential equations.

Equation (1.3) means that the divergence of the electric current density ($\mathbf{J}$) must equal the temporal change of the charge density. This does not seem to help much, since we have already said that ρ is small. In general, $\mathbf{J}_{\text{vec}} = \sigma \mathbf{E}_{\text{vec}} + \rho V_{\text{vec}}$, where V_{vec} is the velocity of the fluid medium. But if ρ is small, then $\mathbf{J} = \sigma \mathbf{E}$, the differential form of Ohm's law, is useful for describing electric fields in space plasmas such as the ionosphere. At the end of this section, we will consider the case of the lower atmosphere where the second term, ρV_{vec}, can exceed the Ohmic current, $\sigma \mathbf{E}$.

Now consider a uniform conducting medium with $\mathbf{J} = \sigma \mathbf{E}$. Then we have, from Eq. (1.3):

$$\nabla \cdot (\sigma \mathbf{E}) = -\frac{\partial \rho}{\partial t}.$$

Suppose σ is uniform. Then consider the effect on the electric field in a medium caused by a free charge q as given above to be $E(r) = Kq/r^2$ (also known as Coulomb's law). This charge q will attract charge in the conducting medium with the opposite sign of charge and will repel the charge of the same polarity as q. Thus, in a small volume around q, a negative charge of $-q$ will quickly accumulate. Therefore, the total electric field will decrease much faster with distance than just $1/r^2$. In fact, within a distance depending on the temperature and number density, the total charge $q + (-q)$ will be a net zero. This happens throughout the conducting medium, thus rapidly forcing the total charge density to be net zero. This process is called screening. Even with large-scale current flows, the total charge density deviates from zero by such a small amount that it is difficult to measure.

Thus, looking back at Eq. (1.3), a good approximation for all times τ longer than it takes for the screening to occur (given by the ratio of the conductivity as $\sigma/\varepsilon_0 = \tau$), the charge density and its time dependence will be essentially zero, so we can write:

$$\nabla \cdot \mathbf{J} = 0. \tag{1.4}$$

This, in turn, says that if some current exists, say, due to a generator of some type, an electric field builds up instantaneously to (almost) force $\nabla \cdot \mathbf{J} = 0$. This is the equation we use to find the steady-state $\mathbf{E}$.

The second electric field generation equation, (1.2), shows that the electric and magnetic fields are closely coupled when time variation is allowed. The third Maxwell equation,

$$\nabla \times \mathbf{B} = \mu_0 \cdot \mathbf{J} + \varepsilon_0 \mu_0 \frac{\partial \mathbf{E}}{\partial t}, \tag{1.5}$$

completes the coupling picture. $\mathbf{B}$ makes $\mathbf{E}$ and $\mathbf{E}$ makes $\mathbf{B}$. This equation can also be written:

$$\nabla \times \mathbf{B} = \mu_0 \left(\mathbf{J} + \varepsilon_0 \frac{\partial \mathbf{E}}{\partial t} \right), \tag{1.6}$$

where $\mathbf{J}$ is called the conduction current and $\varepsilon_0(\partial \mathbf{E}/\partial t)$ is called the displacement current. Equations (1.2) and (1.6) are all we need to explain electromagnetic waves, which permeate space and can also reach the ground. As noted above, wave electric fields are discussed in Chapter 6.

In addition to these fundamental equations, we introduce the concept of continuous media having properties of electrical conduction. This conduction is then described by equations of the form

$$\mathbf{J} = \sigma \mathbf{E}. \tag{1.7}$$

For an isotropic medium such as the lower atmosphere below about 80 km, using an isotropic conductivity σ is completely adequate; however, see the last part of this section, which includes currents caused by the motion of nonzero charge density ρ. The criterion for isotropic conductivity is that collisions between the charged particles (v_{en} and v_{in}) and the neutral air occur at a much higher rate than the particle gyrofrequencies (Ω_i and Ω_e) about the magnetic field, that is, $v_{jn} \gg \Omega_j$. In the atmosphere, then, the electron momentum equation (ignoring gravity) can be written as:

$$nm\frac{dv_e}{dt} = -neE - nmv_e v_{en}.$$

In a steady state, uniform fluid thus has the solution:

$$v_e = -\frac{eE}{mv_{en}}.$$

For heavy ions at rest, the current is then

$$\mathbf{J} = -nev_e = \left(\frac{ne^2}{mv_{en}}\right)\mathbf{E}.$$

Including ions, we have:

$$\sigma = \frac{N_i e^2}{M v_{in}} + \frac{N_e e^2}{m v_{en}}, \tag{1.8}$$

where $m(M)$ is the electron (ion) mass. In clouds, one must take into account charged aerosols, which are very complex and interesting and must be treated separately.

Above 80 km, $\Omega_e > v_{en}$ and the electrons are magnetized, but the ions are not. The atmosphere is exponentially decreasing in density and, by 120 km, the ions are also magnetized Between 80 and 120 km, we have the most complex conductivity model. Taking $\mathbf{B}$ parallel to $\mathbf{a}_z$, we have (see Kelley, 1989, 2009),

$$\mathbf{J} = \sigma\cdot\mathbf{E},$$

where

$$\sigma = \begin{pmatrix} \sigma_P & -\sigma_H & 0 \\ \sigma_H & \sigma_P & 0 \\ 0 & 0 & \sigma_0 \end{pmatrix} \tag{1.9}$$

and

$$\sigma_P = \left[\frac{1}{m\left(1+K_i^2\right)} + \frac{1}{m\left(1+K_e^2\right)} \right] \frac{ne^2}{v}, \tag{1.10a}$$

$$\sigma_H = \left[\frac{\kappa_e^2}{1+\kappa_e^2} - \frac{\kappa_i^2}{1+\kappa_i^2} \right] \frac{ne}{B}, \tag{1.10b}$$

$$\sigma_0 = \left[\frac{1}{Mv_{in}} + \frac{1}{mv_{en}} \right] ne^2, \tag{1.10c}$$

are the Pedersen, Hall, and specific or parallel conductivities, respectively. Here, $\kappa_{i,e}$ is the ratio of Ω/v for each species.

Above 120 km, σ is diagonal since the Hall conductivity vanishes. To a good approximation, in the thermosphere and the collocated ionosphere, e.g., above 150 km,

$$\sigma_P = \frac{ne^2 v_{in}}{M\Omega_i^2}.$$

In a plasma, note that $n_i \approx n_e \approx n$, the plasma density.

To summarize, in the range of 80-120 km, $\mathbf{J}$ has components that are both parallel and perpendicular to $\mathbf{E}$. Above and below this range, $\mathbf{J}$ is everywhere parallel to $\mathbf{E}$.

Above 1000 km, conditions change once again. In general, for $\Omega \gg v$ in response to any force, $\mathbf{F}$, charged particles move across $\mathbf{B}$ according to the equation:

$$\mathbf{V}_D = \frac{\mathbf{F} \times \mathbf{B}}{q|\mathbf{B}|^2}. \tag{1.11}$$

If $\mathbf{F} = q\mathbf{E}$, we have the curious result that

$$\mathbf{V}_D = \frac{\mathbf{E} \times \mathbf{B}}{|\mathbf{B}|^2} \qquad (1.12)$$

is the same for ions and electrons. In a collisionless magnetized plasma, a steady electric field creates no current whatsoever. Ohm's law does not involve $\mathbf{E}$ at all, although it does involve $\partial\mathbf{E}/\partial t$, which is a form of displacement current. Forces of importance include gravity, pressure gradient, inertia, diamagnetism, and curvature effects.

The above discussion can be termed the guiding center approach to magnetized plasmas. We use this concept occasionally but the fluid approach is simpler and often more correct. In this model, the perpendicular plasma velocity is given by Eq. (1.12), where $\mathbf{E}$ and $\mathbf{V}$ must be measured in the same reference frame (see below). Crucial equations in the fluid description include momentum, which is

$$\rho\frac{d\mathbf{V}}{dt} = \rho\mathbf{g} - \nabla p + \mathbf{J} \times \mathbf{B}, \qquad (1.13a)$$

and mass continuity,

$$\frac{\partial\rho}{\partial t} + \nabla\cdot(\rho\mathbf{V}) = 0. \qquad (1.13b)$$

In the lowest part of the atmosphere, where the thunderstorms generate lightning (see Figs. 1.1 and 1.2), charge can move not only by the influence of electric fields (called Ohmic currents) but also when a wind with velocity $\mathbf{V}$ moves a volume of air containing a net charge density ρ. In that case, we must find the total current from

$$\mathbf{J} = \sigma\mathbf{E} + \rho\mathbf{V},$$

where $\mathbf{V}$ is the fluid velocity, in this case, the same as the plasma velocity discussed above because all of the charge carriers are ions (positive and negative), which are carried along by the neutral fluid velocity. In the global electric circuit (discussed in Chapter 2), large electric field gradients are generated below 20-40 km, so our equation for $\nabla \cdot \mathbf{E} = \rho/\varepsilon_0$ tells us that there must be a significant net nonzero charge density, ρ. With significant ρ and some air motion $\mathbf{V}$, there must be a current just from the second term, $\rho\mathbf{V}$. This becomes particularly important as we examine the electric field as a function of altitude, as we will do in Chapter 2.

There is one last important relationship for the lower atmosphere and it comes from looking at the nonzero density, ρ. As for the plasmas discussed above, the charge density is given by $\rho = (Z_i n_i - Z_n n_e)e$ where now we must allow for the possibility that the negative charge carriers are negative ions, each with a Z_n charge. As in the higher altitudes, the number of negative charges, $Z_n n_e$, is nearly equal to the number of positive charges, $Z_i n_i$, so we can speak of the ion density of species s (where s represents either the positive or negative ions). The density of the ions can be changed by ionization, such as when a cosmic ray secondary particle smashes into an atom with enough energy to give one or more bound electrons the energy to escape the atom. This increases the species density.

On the other hand, because positive and negative charges are attracted and there is a strong tendency for positive and negative ions to exchange charge, you wind up with two neutral atoms or molecules. The species charge density is also reduced when an ion attaches to a large object such as an aerosol particle. If that happens, the net mass attached to that charge is so large that it can barely move and has difficulty participating in any current. So, in general, the species density n_s has this time dependence:

$$\mathrm{d}n_s/\mathrm{d}t = \widetilde{q} - \alpha n_s^2 - \beta N_A n_s,$$

where n_s is either the positive or negative ion number density, α is the recombination rate coefficient, and β is the attachment rate coefficient to aerosols with aerosol density N_A. From this equation, one can see that, in the steady state where n_s is not changing, the right-hand side must be equal to zero. Thus, in the case of no aerosols, one can solve for the n_s as

$$n_s = \sqrt{(\tilde{q}/\alpha)}.$$

That is, given the ionization rate $\tilde{q}$, we find that the ion density comes to an equilibrium value that depends on the recombination rate coefficient.

1.4 RELATIVISTIC EFFECTS

Surprisingly, relativistic effects actually matter, even in slowly moving fluids such as the atmosphere. For example, for two frames differing by velocity $\mathbf{V}$ we have:

$$\mathbf{E}' = \mathbf{E} + \mathbf{V} \times \mathbf{B}, \tag{1.14}$$

$$\mathbf{B}' = \mathbf{B}, \tag{1.15}$$

for $|\mathbf{V}| \ll c$, the speed of light.

For the earth's case, if we take $|\mathbf{V}| = 100$ m/s, $|\mathbf{B}| = 0.5 \times 10^{-4}$ T, and $|\mathbf{E}| = 0$, $|\mathbf{E}'| = 5$ mV/m. This may sound small, but 5 V/km is a lot when the medium temperature is 0.1 eV, as it is in the ionosphere. Also, distances are large and, consequently, appreciable potential differences can be involved. We will use (1.14) several times below. Equation (1.15) says that the magnetic field and, by implication, the current density are unchanged by a nonrelativistic reference frame change. We use this fact as well.

1.5 ELECTRIC FIELD MAPPING

The high conductivity parallel to the earth's magnetic field, σ_0, has important implications concerning the transmission of electric fields for long distances along the magnetic field. In fact, if σ_0 were infinite, there would be a zero potential drop along the magnetic field, and the potential difference between any two field lines would be constant. In such a case, any electric field generated at ionospheric heights would be transmitted along the magnetic field lines to very high altitudes. For example, an electric field generated at $60°$ magnetic latitude would be communicated to the equatorial plane at an altitude of over 25,000 km. Likewise, electric fields of solar wind or magnetospheric origin could be transmitted to ionospheric heights.

This phenomenon can be studied quantitatively as follows (following Farley, 1959, 1960). Suppose first that the conductivity is anisotropic but uniform and that the neutral wind is absent. If the electric field perpendicular to **B** is $\mathbf{E}_\perp$ and the field parallel to **B** is $\mathbf{E}_\parallel$, the total current is

$$\mathbf{J} = \sigma_P \mathbf{E}_\perp - \sigma_H (\mathbf{E}_\perp \times \hat{B}) + \sigma_0 \mathbf{E}_\parallel. \tag{1.16}$$

For an electrostatic field, $\mathbf{E} = -\nabla\phi$. Substituting this expression for **E** into (1.16), taking the divergence, setting $\nabla \cdot \mathbf{J} = 0$, and taking **B** to be in the z direction yields

$$-(\sigma_P)\partial^2\phi/\partial x^2 + (\sigma_H)\partial^2\phi/\partial x\partial y - (\sigma_P)\partial^2\phi/\partial y^2,$$

$$-(\sigma_H)\partial^2\phi/\partial y\partial x - (\sigma_0)\partial^2\phi/\partial z^2.$$

The terms containing σ_H cancel, leaving

$$\partial^2\phi/\partial x^2 + \partial^2\phi/\partial y^2 + (\sigma_0/\sigma_P)\partial^2\phi/\partial z^2 = 0. \tag{1.17}$$

Making the change of variables,

$$dz' = (\sigma_P/\sigma_0)^{1/2}dz,$$
$$dx' = dx,$$
$$dy' = dy,$$

converts (1.17) to

$$(\nabla')^2\phi = 0, \tag{1.18}$$

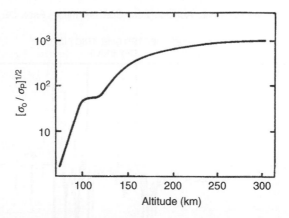

Fig. 1.7 The mapping ratio $(\sigma_0/\sigma_P)^{1/2}$ plotted as a function of height for a typical mid- to high-latitude ionosphere. After Kelley (2009). Reproduced with permission of Elsevier.

which is Laplace's equation in the "reduced" coordinate system. That is, the substitution has transformed the real medium into an equivalent isotropic medium with a greatly reduced depth parallel to the magnetic field (the z direction in the calculation). The ratio $(\sigma_0/\sigma_P)^{1/2}$ is plotted in Fig. 1.7 for a typical ionospheric profile. Above 130 km, the ratio exceeds 100, reaching 1000 at 300 km. At high altitudes, σ_0 becomes independent of density. The ratio σ_0/σ_P continues to increase as both the ion-neutral collision frequency and the plasma density, which determine σ_P, continue to decrease. One of the basic approximations of MHD is that if the conductivity parallel to the magnetic field becomes very large, the parallel electric field component vanishes. For many applications, MHD theory applies on the high-altitude portions of the field lines that contact the ionosphere.

The implication of these calculations is that electrical features perpendicular to **B** map for long distances along the earth's magnetic field lines. This has been verified experimentally via the simultaneous measurements shown in Fig. 1.8. In this experiment, the zonal electric field component was measured in the Northern Hemispheric ionosphere with a radar (see Appendix A of Kelley (2009), where a number of

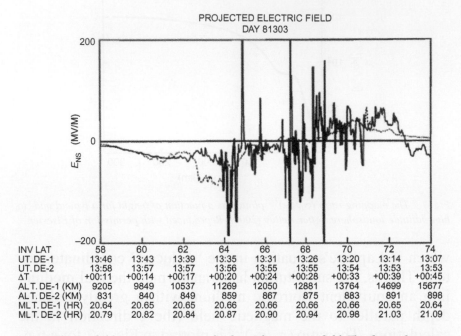

WEIMER ET AL.: AURORAL ZONE ELECTRIC FIELDS FROM DE 1 AND 2

PROJECTED ELECTRIC FIELD
DAY 81303

INV LAT	58	60	62	64	66	68	70	72	74
UT. DE-1	13:46	13:43	13:39	13:35	13:31	13:26	13:20	13:14	13:07
UT. DE-2	13:58	13:57	13:57	13:56	13:55	13:55	13:54	13:53	13:53
ΔT	+00:11	+00:14	+00:17	+00:20	+00:24	+00:28	+00:33	+00:39	+00:45
ALT. DE-1 (KM)	9205	9849	10537	11269	12050	12881	13764	14699	15677
ALT. DE-2 (KM)	831	840	849	859	867	875	883	891	898
MLT. DE-1 (HR)	20.64	20.65	20.65	20.66	20.66	20.66	20.66	20.65	20.64
MLT. DE-2 (HR)	20.79	20.82	20.84	20.87	20.90	20.94	20.98	21.03	21.09

Fig. 1.8 Electric field components perpendicular to the magnetic field. This figure shows two satellites at altitudes between 840 and over 10,000 km. The electric field, when mapped along magnetic field lines to the ionosphere, is nearly identical.

measurement techniques are discussed) and in the inner magnetosphere at a point very close to where the same magnetic field line crossed the equatorial plane. The latter measurement was accomplished using the whistler technique (Carpenter et al., 1972). The two measurements clearly have the same temporal form, but the magnetospheric component is 10 times smaller. This difference may be explained as a geometric effect arising from the spreading of magnetic field lines as follows. First, we take σ_0 to be infinite so that the field lines act like perfect conductors. This implies that no potential difference exists along them and, in turn, that the voltage difference between two lines is conserved. The magnetic field decreases along the field line as the square of the distance from the mid- or high-latitude ionosphere to the equatorial plane in the magnetosphere. Since the voltage between adjacent field lines is constant, the perpendicular electric field must also decrease along the field lines linearly and E/B is constant. For a dipole field, Mozer (1970) showed that the two electric field components (meridional E_{MI} and zonal E_{ZI}) map from the ionosphere to the magnetosphere in the equatorial plane as

$$E_{MI} = 2L(L - 3/4)E_{RM}, \qquad (1.19a)$$

$$E_{ZI} = L^{3/2}E_{ZM}, \qquad (1.19b)$$

where the L value is the distance from the center of the earth to the equatorial crossing point measured in earth radii (R_e), E_{RM} is the radial magnetospheric component at the equatorial plane, and E_{ZM} is the zonal magnetospheric component there. The equation for the zonal component (1.19b) is in excellent agreement with the corresponding data in Fig. 1.8. Notice that the zonal ionospheric electric field component maps to a zonal field in the equatorial plane but that the meridional component in the ionosphere, E_{MI}, becomes radial in the equatorial plane (E_{RM}). In particular, a poleward ionospheric electric field points radially outward at the equatorial plane. Thus, large-scale electric fields generated in the E and F regions of the ionosphere can map upward to the magnetosphere and create motions there. Likewise, electric fields of magnetospheric and solar wind origin can map from deep space to ionospheric heights and have even been detected by balloons at stratospheric heights (Mozer and Serlin, 1969) (see Chapter 2).

Farley (1959) studied the upward mapping process realistically by including the z dependence of the conductivities in his analysis. The basic equations, $\nabla \cdot \mathbf{J} = 0$, $\mathbf{E} = -\nabla\phi$, and $\mathbf{J} = \sigma \cdot \mathbf{E}$, are the same ones used in deriving (1.18) and, with the assumption that variations of σ occur only in the z direction, they yield

$$\partial^2\phi/\partial x^2 + \partial^2\phi/\partial y^2 + (1/\sigma_P) + \partial/\partial z(\sigma_0 \partial\phi/\partial z) = 0.$$

The same change of variables now yields

$$(\nabla')^2\phi + (\partial\phi/\partial z')(\partial/\partial z')\left[\ln(\sigma_0\sigma_P)^{1/2}\right] = 0, \qquad (1.20a)$$

where $(\sigma_0\sigma_P)^{1/2} = \sigma_m$ is termed the geometric mean conductivity. Furthermore, if σ_m can be modeled in the form $\sigma_m = c \exp(c_0 z')$, then the equation simplifies to

$$(\nabla')^2\phi + c_0(\partial\phi/\partial z') = 0. \tag{1.20b}$$

This differential equation has a straightforward analytical solution. By considering the solutions with different Fourier wave numbers in the source field, Farley showed that (a) larger-scale features map more efficiently to the upper thermosphere (ionospheric F region) from the lower thermosphere (ionospheric E region) than small-scale features do; (b) the source field height is very important, with upper E-region structures highly favored as F-region sources; and (c) roughly speaking, perpendicular structures with scale sizes greater than a few kilometers map unattenuated to F-region heights. The implication here is that if very large-scale electric fields are generated in the E region, the resulting potential differences map up into the F-region ionosphere and beyond along the magnetic field lines, deep into space. As we shall see, these low-altitude electric fields dominate motions of the plasma throughout the dense plasma region around the earth, termed the plasmasphere.

For example, considering sources at the F region and at even higher altitudes in the magnetosphere and solar wind, the previous analysis can be used to show that the mapping efficiency to the E region is huge. In fact, within the magnetospheric and solar wind plasmas, the parallel conductivity is often taken to be infinite and hence the parallel electric field vanishes, even when finite field-aligned currents flow. As we shall see later, this assumption that $E_{\parallel} = 0$ is a powerful analytical device that allows great conceptual simplifications in the understanding of magnetospheric electric field and flow patterns. On the other hand, in the regions where the assumption of infinite conductivity breaks down is exactly where very interesting phenomena occur. Generation of the aurora is an example, as is magnetic connection and reconnection.

1.6 AN ENERGY THEOREM

Before concluding this section, it is useful to derive an energy relationship based on MHD principles. The particle pressure, $p = nk_BT$, where k_B is the Boltzmann's constant, is equivalent to the particle energy density, and analogies occur for magnetic pressure and magnetic energy density; that is, $B^2/2\mu_0$ yields the energy stored in a magnetic field per unit volume as well as the magnetic pressure. The total stored magnetic energy in a system is then

$$W_B = (1/2\mu_0)\int B^2 dV,$$

where dV is the volume element, and we have used a single integral sign to designate a triple integral. In the magnetosphere, the electric field energy density is tiny compared to the magnetic energy density. Changes of this quantity with time can be written as

$$\partial W_B/\partial t = (1/\mu_0)\int (B \cdot \partial B/\partial t)dV.$$

Then, using $\partial \mathbf{B}/\partial t = -\nabla \times \mathbf{E}$, $\mathbf{B} = \mu_0 \mathbf{H}$, and the vector identity $\nabla \cdot (\mathbf{E} \times \mathbf{B}) = \mathbf{B} \cdot (\nabla \times \mathbf{E}) - \mathbf{E} \cdot (\nabla \times \mathbf{B})$, we have

$$\partial W_B/\partial t = -\int \nabla \cdot (\mathbf{E} \times \mathbf{H})d\Omega - \int \mathbf{E} \cdot (\nabla \times \mathbf{H})dV.$$

Using $\mathbf{J} = \sigma(\mathbf{E} + \mathbf{V} \times \mathbf{B})$ and $\mathbf{J} = \nabla \times \mathbf{H}$ and applying the divergence theorem to convert the first volume integral to a surface integral gives

$$\partial W_B/\partial t = -\iint_\Sigma (\mathbf{E} \times \mathbf{H}) \cdot d\mathbf{a} - \int [(\mathbf{J}/\sigma - \mathbf{V} \times \mathbf{B}) \cdot \mathbf{J}]dV,$$

where the area element, $d\mathbf{a}$, points outward from the surface of the volume. Finally, rearranging this equation yields

$$\partial W_B/\partial t = -\iint_\Sigma (\mathbf{E} \times \mathbf{H}) \cdot d\mathbf{a}$$
$$-\int (J^2/\sigma)dV - \int [\mathbf{V} \cdot (\mathbf{J} \times \mathbf{B})]dV. \qquad (1.21)$$

In words, the change in stored magnetic energy in a volume equals the energy flux across the surface into the volume in the form of the Poynting flux ($\mathbf{E} \times \mathbf{H}$), minus the resistive energy loss inside the volume, minus the mechanical work done against the $\mathbf{J} \times \mathbf{B}$ force inside the volume.

In a truly closed magnetosphere with the surface an equipotential, no energy crosses the surface ($\mathbf{E}$ is everywhere normal to an equipotential) and there could be no internal circulation (convection), no dissipation by ionospheric currents, and no storage of magnetic energy for later release (e.g., in auroral substorms). Two sources for generating a component of $\mathbf{E}$ parallel to the magnetopause, and hence a net Poynting flux inward, are viscous interaction and reconnection. Both of these processes thus result in a net flow of energy into the magnetosphere and are discussed in Chapter 3.

1.7 SUMMARY

In this chapter, a brief introduction to electric field generation and mapping was presented, including relativistic effects. The conductivity tensor was introduced and Poynting's theorem was discussed.

Finally, it has been argued by Vasyliūnas (2012) that the electric field is not "fundamental" and that it arises only because of plasma motion. This is easy to understand in a collisionless plasma such as the solar wind. Due to the high conductivity, there is no electric field in a reference frame moving with the plasma. But in the earth's reference frame, there is an electric field. It drives the aurora and affects the temperature and neutral wind in the thermosphere. The four generators discussed in Chapters 2–5 all compete to determine the earth's electric field.

REFERENCES

Boeck, W.L., Vaughan Jr., O.H., Blakeslee, R., Vonnegut, B., Brook, M., 1992. Lightning induced brightening in the airglow layer. Geophys. Res. Lett. 19 (2), 99–102.

Briggs, M.S., et al., 2010. First results on terrestrial gamma ray flashes from the Fermi Gamma-ray Burst Monitor. J. Geophys. Res. 115 (A7), A07323. http://dx.doi.org/10.1029/2009JA015242.

Briggs, M.S., et al., 2011. Electron-positron beams from terrestrial lightning observed with Fermi GBM. Geophys. Res. Lett. 38 (2), L02808. http://dx.doi.org/10.1029/2010GL046259.

Carpenter, D.L., Stone, K., Siren, J.C., Crystal, T.L., 1972. Magnetospheric electric fields deduced from drifting whistler paths. J. Geophys. Res. 77 (16), 2819–2834.

Connaughton, V., et al., 2013. Radio signals from electron beams in terrestrial gamma ray flashes. J. Geophys. Res. Space Phys. 118 (1–8), 2313–2320. http://dx.doi.org/10.1029/2012JA018288.

Craven, J.D., Frank, L.A., 1991. Diagnosis of auroral dynamics using global auroral imaging with emphasis on large-scale evolution. In: Meng, C.-I., Rycroft, M.J., Frank, L.A. (Eds.), Auroral Physics. Cambridge University Press, Cambridge, UK, pp. 273–288.

Farley, D.T., 1959. A theory of electrostatic fields in a horizontally stratified ionosphere subject to a vertical magnetic field. J. Geophys. Res. 64 (9), 1225–1233.

Farley, D.T., 1960. A theory of electrostatic fields in the ionosphere at nonpolar geomagnetic fields. J. Geophys. Res. 65 (3), 869–877.

Franz, R.C., Nemzek, R.J., Winckler, J.R., 1990. Television image of a large upward electrical discharge above a thunderstorm system. Science. 249 (4964), 48–51. http://dx.doi.org/10.1126/science.249.4964.48.

Gerken, E.A., Inan, U.S., Barrington-Leigh, C.P., 2000. Telescopic imaging of sprites. Geophys. Res. Lett. 27 (17), 2637–2640.

Kelley, M.C., 1989. The Earth's Ionosphere: Plasma Physics and Electrodynamics, International Geophysics Series, vol. 43. Academic Press, San Diego, CA.

Kelley, M.C., 2009. The Earth's Ionosphere: Plasma Physics and Electrodynamics, second ed. International Geophysics Series, vol. 96. Academic Press, Burlington, MA.

Lyons, W.A., et al., 1994. Characteristics of luminous structures in the stratosphere above thunderstorms as imaged by low-light video. Geophys. Res. Lett. 21 (10), 875–878.

Mozer, F.S., 1970. Electric field mapping in the ionosphere at the equatorial plane. Planet. Space Sci. 18 (2), 259–263.

Mozer, F.S., Serlin, R., 1969. Magnetospheric electric field measurements with balloons. J. Geophys. Res. 74 (19), 4739–4754.

Rakov, V.A., Uman, M.A., 2003. Lightning: Physics and Effects. Cambridge University Press, Cambridge, UK.

Vasyliūnas, V.M., 2012. The physical basis of ionospheric electrodynamics. Ann. Geophys. 30, 357–369. http://dx.doi.org/10.5194/angeo-30-357-2012.

CHAPTER 2

Atmospheric Electricity

The Earth's Electric Field. http://dx.doi.org/10.1016/B978-0-12-397886-8.00002-8

Below 80 km, the atmosphere is not plasma in the usual sense because charged particle motion is controlled more closely by neutral drag than by large-scale electromagnetic forces. Nevertheless, the atmosphere is a weak, isotropic electrical conductor with a myriad of electrodynamic phenomena. This conductivity of the air becomes weaker and weaker approaching the earth. Compared to the air at sea level, the earth is essentially a perfect conductor. Negative cloud-to-ground (CG) lightning charges the earth negatively, creating a worldwide "fair weather" electric field pointing vertically downward and continually acts to discharge the planet.

We discuss this global circuit in this chapter, as well as the elements driving it from which secondary phenomena are produced with effects in the upper atmosphere, ionosphere, and magnetosphere. The focus of this chapter will be on the lower atmosphere, which is dominated by thunderstorm-generated charge sources, but we will also discuss the downward coupling of magnetospheric electric fields into the atmosphere, which can dominate electrical phenomena in fair weather for some vector directions.

2.1 THE FAIR WEATHER ELECTRIC FIELD

At about the time when Ben Franklin was teaching about the electrical nature of lightning (Chapter 1), Le Monnier (1752) first demonstrated that the atmosphere was electrified, even in fair weather. As it became understood that a conducting fluid existed in the air, Linss (1887) first realized that this conducting fluid would soon cancel out the field produced by any bound charge. Long before Elster and Geitel (1899a,b) and Wilson (1900) discovered small ions in the air, Coulomb (1795) found that the air itself was conducting and that charge on a conductor would leak away in the air. Thus, two unsolved mysteries, known since the late 1700s, remain about fair weather electricity: (1) What was the source of the large-scale electric field and why was it not quickly reduced to zero by charges in the conducting fluid? (2) What was the source of the conducting fluid or, in other words, why didn't the large-scale electric field sweep away all the free charge in the atmosphere? Another critical discovery came in 1911 when Victor Hess was conducting an atmospheric electricity experiment to study radioactive ions in the air. Using a balloon experiment, he found that the conductivity increased with altitude rather than the expected decrease, thus leading to the discovery of cosmic radiation (Hess, 1911). Putting these ideas together, Wilson (1920) first proposed that the earth was a conductor and that thunderstorms over the globe must act to put negative charge on the ground and positive charge in the upper atmosphere, functioning like a leaky capacitor. Thus, the concept of a global electric circuit was born (see Fig. 2.1).

The idea of a global circuit was quickly accepted by the scientific community after Mauchly (1923) used electric field data from the Carnegie research vessel to show that the fair weather electric field seemed to have a peak at around 19:00 h UT,

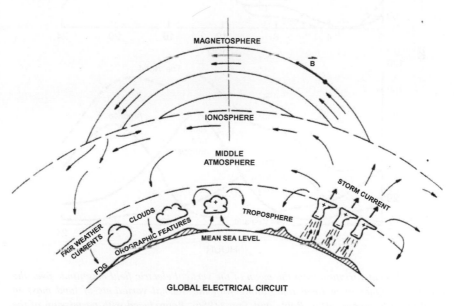

Fig. 2.1 Schematic of the global electrical circuit. After Roble and Tzur (1986). Reproduced with permission of the National Academy of Sciences.

independent of where the measurements were taken over the globe. Then, Whipple and Scrase (1936) connected these global field variations with the fact that thunderstorms occur most frequently over land in the afternoon (local time) and proposed that the global circuit daily variation in fair weather was simply caused by variation of the global land mass as the earth rotates (see Fig. 2.2). At 19:00 UT, there is more land mass in the late afternoon than during other times of the day, such as 04:00 UT when the Pacific Ocean is in late afternoon (local time) and the global circuit has a minimum.

Thus, for more than a century, atmospheric electrodynamics has been known to be intimately connected to global phenomena in the upper atmosphere and space (Chalmers, 1967).

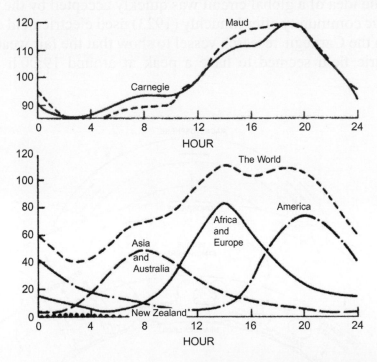

Fig. 2.2 Percent variation from the mean of the vertical electric field magnitude over the ocean as a function of local time, along with the diurnal variations of land mass in local-time afternoons. After Roble and Tzur (1986). Reproduced with permission of the National Academy of Sciences.

A simple model of the global electric current system is as follows. Consider the earth and the ionosphere to be charged and functioning like capacitor "plates" enclosing a conducting medium. If a source was not charging the earth, the capacitor would simply discharge due to the conducting medium like a leaky capacitor. On the other hand, if no method existed to create new ions between the plates, the potential difference would soon sweep out the charge between the plates, shutting off the global current. The source of the ions is dominated by galactic cosmic rays with energies typically 100 GeV or more, each of which slams into the atmosphere, leaving a trail of highly ionized gas in its wake. These energetic charged tracks left by the cosmic rays cause extensive air showers (EAS) of relativistic secondaries at the earth's surface, and most of the millions of electron-ion pairs produced in the process immediately recombine. However, a small percentage of the free charge of the EAS attaches to air molecules, becoming much less mobile, and therefore remains in the air and disperses after most of the EAS charge has recombined. These residual ions form the basis of the atmospheric conductivity at all altitudes above about 1 km over solid ground (below that, atmospheric ions are mostly produced by the radioactive decay of natural outgassing from the earth, such as ^{222}Rn). The ion pair production rate as a function of altitude is shown in Fig. 2.3, where we see that the peak production rate occurs between 10 and 30 km. The rate decreases above this layer because the lower density air in the upper atmosphere presents less of a target for cosmic rays. Below 10 km, most cosmic rays and their secondary particles have already spent their energy, leaving fewer total ion pairs at the lowest levels.

The average conductivity created by these ions is plotted versus altitude in Fig. 2.4 and is nearly exponential below 80 km (Hale, 1984). We see in this figure that the conductivity still increases above the pair production rate peak shown in Fig. 2.3 because the ion-neutral collision frequency drops dramatically with increasing altitude, thus increasing the

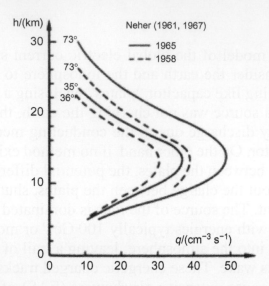

Fig. 2.3 *Pair production rate plot. Profiles of the ionization rate at different latitudes in years of the minimum (1965) and maximum (1958) of the 11-year solar sunspot cycle (Neher, 1961, 1967). After Gringel et al. (1986). Reproduced with permission of the National Academy of Sciences.*

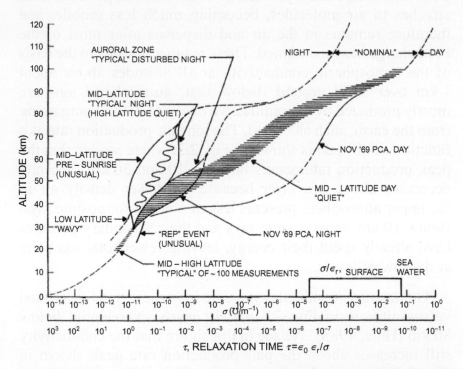

Fig. 2.4 *Electrical conductivity of the atmosphere as a function of altitude. The variability is remarkable, both in time and space, depending on the particle flux of various types. From the ground to the edge of space, σ varies by 10 orders of magnitude. After Hale (1984). Reproduced with permission of the American Geophysical Union.*

conductivity (see Eq. 1.8). Volland (1984) represents the atmospheric resistivity ρ (equal to the inverse conductivity σ) below 60 km altitude with the function of z (km):

$$\rho(z) = \frac{1}{\sigma(z)}$$
$$= \rho_1 \exp(-4.527z) + \rho_2 \exp(0.375z) + \rho_3 \exp(0.121z)$$

where

$$\rho_1 \left(10^{12} \Omega m\right) = 46.9, \quad \rho_2 \left(10^{12} \Omega m\right) = 22.2, \quad \rho_3 \left(10^{12} \Omega m\right) = 5.9.$$

At the earth's surface, the conductivity $\sigma(0) = 1/\rho(0)$ is about 10^{-14} S/m, which is still finite but extremely low. Integrating the column resistivity from the ground to the ionosphere gives 1.2×10^{17} Ωm^2. However, if you divide this number by the earth's surface area, you find that the total resistance from the ground to the ionosphere is only about 230 Ω.

In our simple model, the capacity of a spherical capacitor can be calculated as follows. We first calculate the potential between two concentric spheres for which the outer sphere of radius b is grounded and the inner plate has charge Q. The electric field is then

$$\mathbf{E} = \frac{Q}{4\pi \varepsilon r^3} \mathbf{r}$$

where $\mathbf{r}$ is the radial vector. For air, $\varepsilon \approx \varepsilon_0$, the permittivity of free space. Using $\mathbf{E} = -\nabla V$ where V is the voltage and converting this to an integral,

$$V = -\int_a^b \mathbf{E} \cdot d\mathbf{r} = -\int_b^a \frac{Q}{4\pi \varepsilon_0 r^2} dr.$$

Performing this integral,

$$V = \frac{Q}{4\pi\varepsilon_0}\frac{(b-a)}{ab}.$$

Since $Q = CV$,

$$C = 4\pi\varepsilon_0\frac{ab}{b-a}. \tag{2.1}$$

For our case, the distance $b - a$ is equal to the atmospheric scale height, H, since the conductivity increases so rapidly that the effective height of the outer plate is approximately at the $1/e$ point for resistivity or $H = 7.9$ km. We obtain a capacity of about 0.6 F. Using the value of 230 Ω to approximate the resistance between the plates, we estimate that the time constant for the global circuit should be on the order of 138 s. Estimates of this time constant by others vary from a few minutes to 40 min.

The total voltage between the earth and the ionosphere is difficult to measure instantaneously, but many groups have estimated the voltage to be 250-350 kV using a variety of techniques. Experimentally, we know that the earth carries a net negative charge compared to the upper atmosphere/ionosphere. Direct voltage drops on the order of 200 kV have been measured between sea level and 1.5 km (20% of the 7.9 km exponential scale height discussed above) (Woosley and Holzworth, 1987), representing a charge of $Q = CV \sim 2 \times 10^5$ C. Discharging this amount of charge in an RC time constant results in a current estimate for the global circuit of $I \sim 1500$ A. Thus, $\sim 10^3$ A is the order of magnitude for the sum of all the global thunderstorms in series to drive the global circuit. If all of these storms turned off suddenly, the earth would completely discharge within 10-30 min. Bering et al. (1998) discussed estimates for the global circuit elements, noting that the most common type of lightning stroke to ground puts negative charge on the earth. However, there remains a controversy as to whether the dominant charging mechanism results from the net lightning charge to ground or is more related to the quasi-dc current from the whole thunderstorm. Many measurements over thunderstorms suggest that each storm produces between 0.1 and 10 A of continuous, upward vertical current (positive charge moving upward). Thus, the $\sim 10^3$ A in the global circuit could be produced by the $\sim 10^3$ thunderstorms estimated to be active simultaneously.

2.2 INTRODUCTION TO CLOUD CHARGING

Figure 2.5 shows the distribution of total annual lightning activity for the earth. This figure demonstrates that the continents are clearly the major source of activity, with comparatively little total lightning over the oceans. This distribution provides a strong hint as to the source of thunderstorm activity: convective activity driven by late afternoon solar heating of the earth's surface. Parcels of warm, moist air rise due to buoyancy and, as the

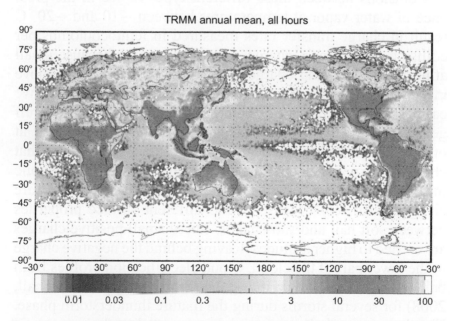

Fig. 2.5 Average global lightning activity recorded from the NASA Tropical Rainfall Measurements Mission (TRMM) LIS (Lightning Imaging Sensor). (For a color version of this figure, the reader is referred to the online version of this chapter.) Figure courtesy of NASA.

air rises, it cools until the moisture starts to condense, thus giving off the thermal energy of the water vapor (latent heat of condensation). This keeps the rising air warmer than the surrounding air. As the air continues to rise, it continues to cool and, eventually, the water droplets start to freeze. As the water freezes, it gives off even more latent heat, again keeping rising air warmer than it would otherwise be, creating a strong updraft. Ice can form by many avenues, such as snowflake crystals by direct vapor deposition or by the aggregation of supercooled water drops as they collide with each other. One of the most amazing properties of water, unlike most substances, is that it expands when it freezes. Thus, because of surface tension, water drops are kept from freezing when they reach $0\,°C$. These supercooled drops accrete onto small hail particles, forming another type of ice with surface properties that are very different from the vapor-grown snowflakes.

Collisions between these different types of ice in the presence of water vapor at temperatures between -10 and $-20\,°C$ result in charge transfer rates measured in femtocoulombs per collision. Large-scale charge separation results because small atmospheric ions and charged ice of different mass can move in different directions and/or speeds within the updraft. We can visualize the updraft as being analogous to a Van de Graff generator, where charge is deposited on the conveyor belt at one point and the mechanical energy of the motor carries that charge to the upper electrode. In a thunderstorm, the most mobile of the charged ice particles are the snowflakes, which are charged positively because of the different surface properties between soft hail and snowflakes. These small positive charges are carried to higher altitudes, leaving behind the negative charge in the lower part of the cloud. Cloud electrification profiles from balloon flights are shown in Fig 2.6 (Stolzenburg and Marshall, 2008) for several storms during the mature thunderstorm phase. The three general charging regions are noted, including a main positive charge near the -25 to $-35\,°C$ isotherms (7-8 km) and, below that, the main (net) negative charge region between

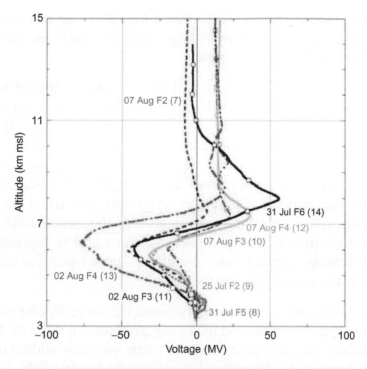

Fig. 2.6 Six cloud electrification profiles through four different mature, main phase thunder-storms. The cloud bases are around 3.7 km altitude, which is close to the +5 to 0 °C isotherm where one typically finds a small positive charge region. This is followed upward by the main negative region near the −15 °C isotherm, which rapidly swings over to the upper positive charge region above 6.5 km altitude and −30° isotherm. (For a color version of this figure, the reader is referred to the online version of this chapter.) Adapted from Stolzenburg and Marshall (2008). Reproduced with permission of the American Geophysical Union.

about −5 and −20 °C, with a small positive charge at the bottom of the cloud. The upper positive region attracts negative charge from the small ions in the clear air above the cloud which, in turn, attach to cloud droplets at the upper cloud boundary. A similar screening effect occurs below the cloud, but it is much less effective because the updraft constantly perturbs the cloud base boundary and atmospheric ions are less mobile at lower altitudes. Near the earth, the high electric field causes corona from sharp points in the terrain (Krehbiel, 1986). This electrification continues until the electrical stress is simply too high. Fields as high as 10 kV/m are commonly observed in and below clouds

(Moore and Vonnegut, 1977). From Fig. 2.6, one can see that these large fields exist over large regions, such as 100 MV between 6 and 7.5 km altitude.

Above the cloud, the electric field is upward, which drives a current upward in the conducting atmosphere, lending a net positive charge to the upper atmosphere and ionosphere. Below the cloud, the average direction of the electric field is also upward, pointing up toward the main negative cloud charge. Between 600 and 1000 simultaneous global thunderstorms convert mechanical (updraft) energy into electrical energy, charging the capacitor and powering the worldwide fair weather electric field, which is downward and is represented by the leakage current from the global, upper positive charge down to the negatively charged earth.

Most lightning is intracloud between the large dipolar charge regions (upper positive and lower negative seen in Fig 2.6). Most CG flashes bring negative charge to earth, violently adding to the charge brought by the upward quasi-steady electric field. When wind shear creates anvil-type clouds, however, the upper positive region can be horizontally displaced as shown in Fig. 2.7 (Krehbiel, 1986). The positive CG strikes occur with corresponding downward current from the base of the ionosphere.

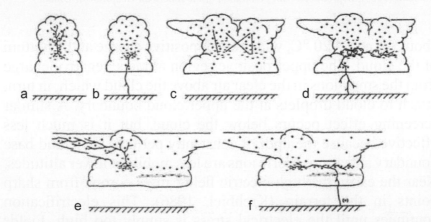

Fig. 2.7 Development of a positively charged anvil cloud. After Krehbiel (1986). Reproduced with permission of the National Academy of Sciences.

2.3 THE LIGHTNING DISCHARGE

Whole books have been written on this topic, such as the ones by Uman (1969, 1987) and Rakov and Uman (2003), and only a brief discussion can be made here.

The most common type of CG stroke brings negative charge to earth. The very first in a series of these CG strokes is now thought to start very high in the cloud with a very intense compact intercloud discharge (CID), also called a narrow bipolar event. This CID is one of the most intense electromagnetic pulses in all of lightning and can even result in bremsstrahlung X-rays so intense that their decay produces positrons (antimatter) as well as electrons (Briggs et al., 2011; Connaughton et al., 2013).

The discharge process then progresses in steps about 50 m long, which is called the stepped leader process wherein ions pool at the tip while new steps of negative charge shoot ahead, forming the next step. Each step brings negative charge closer to the ground. When the tip of the stepped leader nears the ground, a process that takes hundreds of milliseconds, a huge electric field exists near the tip. One can visualize this as being similar to a charged spherical conductor with a narrow wire protruding out from the surface. To maintain constant potential everywhere on the combined conductor, a huge amount of the total charge will be distributed along the narrow wire, just like a lightning streamer protruding down from the large negative charge region in the cloud.

As the streamer tip nears the ground, its intense electric field stimulates upward positive streamers from trees, buildings, etc. The upward streamers attach to the downward negative leader at a few hundred meters altitude and a bright discharge occurs, during which the negative charge on the leader is rapidly transferred to ground, a process referred to as the first-return stroke.

A shock front following the return stroke, initializing the attachment process, moves upward at a speed of 0.1-0.5c ($c =$ speed of light), carrying up to hundreds of kiloamperes

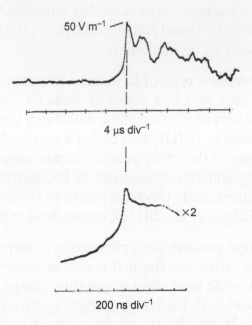

Fig. 2.8 Electric field changes under a discharge. After Krider (1986). Reproduced with permission of the National Academy of Sciences.

for tens of milliseconds, sometimes followed by continuing currents of a few kiloamperes for over a second. About 50 ms later, a new dart leader may follow the same path and the second return strike occurs. Several strokes accompany each flash, discharging different portions of the cloud. Figure 2.8 shows the electric field change under a storm during a first-return lightning strike using two time scales (Krider, 1986). Only a small fraction of the total charge maintaining the 100 MV cloud potential difference discussed above is discharged in the return stroke (note in these plots that a positive E is downward).

2.4 RADIATION FROM LIGHTNING

The short intense current in any of the possible types of lightning strokes radiates similar to that of a dipole antenna or many dipole antennas in series. Antenna theory is well developed for dipole antennas. Over a conducting earth, a mirror effect occurs. If we have a current I_0 over a distance l, the electromagnetic fields are given by

$$H = \frac{I_0 l k^2}{4\pi} \sin\theta \left[\frac{j}{kr} + \frac{1}{(kr)^2} \right] e^{-jkr} \cdot a_\varphi, \qquad (2.2)$$

$$E = \frac{\eta I_0 l k^2}{4\pi} \cos\theta \left[\frac{2}{(kr)^2} - \frac{2j}{(kr)^3} \right] e^{-jkr} a_r,$$

and

$$\frac{\eta I_0 l k^2}{4\pi} \sin\theta \left[\frac{j}{kr} + \frac{1}{(kr)^2} - \frac{j}{(kr)^3} \right] e^{-jkr} a_\theta. \qquad (2.3)$$

The electric field has three terms. The third is the electrostatic field due to the perturbation dipole suddenly created by the lowering of several coulombs of negative charge to ground. This is a dipole-like field due to the positive image charge below. It falls off as r^{-3} and is upward directly above the discharge. The second term is the storage field around the antenna. The first term falls off as r^{-1} and is responsible for the far-field radiation of this channel. This radiation propagates in the earth-ionosphere waveguide all over the planet and is the primary source of noise in the AM radio band. Some of this energy couples into the magnetosphere where it plays an important role in controlling the Van Allen radiation belts. In the magnetic field expression, there is no r^{-3} term since there are no magnetostatic charges.

2.5 COUPLING TO THE IONOSPHERE

In Chapter 6, considerable time is spent describing electromagnetic waves in a magnetized plasma. Without a magnetic field, waves from lightning would either be reflected from the ionosphere or passed into it, depending on whether their frequency was greater or smaller than the peak ionospheric plasma frequency. This frequency is determined by the Fourier components of the lightning source current, $I(t)$, discussed above. Looking at Fig. 2.8 above, in a discharge it is clear that periods from 25 ms (40 kHz) to 100 ns (10 MHz) are present. The peak intensity occurs in the VLF frequency region, around 10 kHz, and a power law of the form f^{-1} exists for $\mathbf{E}(f)$ at higher frequencies. These waves near the peak intensity propagate with little attenuation through very long distances in the earth-ionosphere wave guide, as the wave reflects between the ionosphere and the ground. This phenomenon allows lightning locations to be triangulated anywhere in the world with a relatively small number of receiving stations (see Fig. 1.2). The power spectrum thus falls off as f^{-2}. The measurements in Fig. 2.8 do not reveal the very lowest radiated frequencies, which extend down to a few hertz. Indeed, the so-called Schumann resonances of the earth-ionosphere cavity are excited by lightning at harmonics of the fundamental near 7 Hz.

In the 80-100 km height range, especially during daytime, these waves are more strongly absorbed by collisions of free electrons with the neutrals that occur at a rate corresponding to the dominant VLF frequencies in the lightning waveform. At night, when there is little free electron density in the D region, these electromagnetic waves propagate to higher altitudes with little attenuation before they convert into whistler waves. A most important effect is the coupling from a free space mode to the whistler mode (Helliwell, 1965; Stix, 1962). Rocket measurements over a 3-kHz transmitter in Antarctica show that only about a 10-15-dB signal loss occurs, 3 dB of which is simply because the whistler mode is circularly polarized, whereas the source is linear. Another important effect, which might be called self-absorption, occurs in the D region. The lightning electromagnetic pulse heats the medium, which increases the ion mobility and raises the conductivity above some altitude. Below this altitude, the lightning pulse increases the attachment rates and thus lowers the conductivity at lower D-region altitudes. The resulting heating affects the

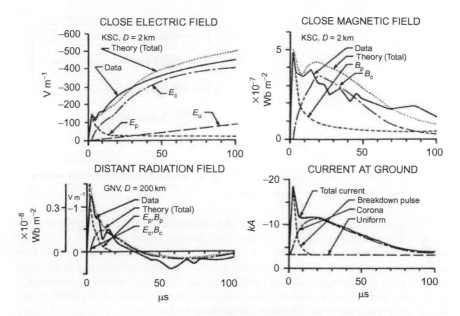

Fig. 2.9 Typical ground-level observations a similar distance away from the storm (bottom left). After Lin et al. (1980). Reproduced with permission of the American Geophysical Union.

ELF phenomenon discussed below. The waves are reflected more effectively or are severely attenuated in this heating effect on conductivity. Indeed, rocket observations 200 km above a lightning discharge reveal fields that are 5-50 mV/m in magnitude. But, as shown in Fig. 2.9 (Lin et al., 1980), by 200 km away horizontally along the earth's surface, the typical field from lightning is 1 V/m from an intense lightning stroke.

A pair of examples from rocket observations is given in Fig. 2.10 in which it is clear from the wave amplitudes (compare to Fig. 2.9) that less than 0.01% of the power is coupled upward into the thermosphere (Fig. 2.10a from Kelley et al., 2004) and that the passage through the ionosphere results in dramatically increased dispersion (Fig. 2.10b from Holzworth et al., 1999). Notice in Fig. 2.10a that there is an electric field component parallel to $\mathbf{B}_0$, something unusual in magnetized plasmas. Such a field will continue to heat electrons and perhaps even accelerate them to high energy. Although there have been no reports of

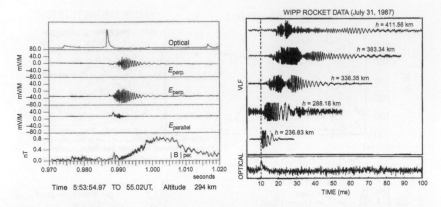

*Fig. 2.10 Rocket electric field measurements above a thunderstorm. (a) The optical and vector electric field and perpendicular magnetic field detected at 294 km altitude over a thunderstorm (after Kelley et al., 1990). (b) The perpendicular-to-**B** axis electric field from several different lightning strokes, all aligned with their detected optical pulse at the rocket (sample given in the bottom panel of b). One can clearly see the increasing dispersion of these waves as the rocket ascends to higher altitudes. After Holzworth et al. (1999). Reproduced with permission of the American Geophysical Union.*

such monopolar pulses above 200 km, at least for the E region, we can use the parallel field shown in Fig. 2.10a to calculate the electron velocity change as

$$\delta v = \frac{e}{m} \int E(\tau) d\tau.$$

This yields about 2 eV. Such a beam may excite plasma waves, which would modify the physical properties of the medium. It certainly would heat the plasma and possibly make detectable airglow. These ideas were clarified by Vlasov and Kelley (2010). However, the monopolar pulse has only been detected below 200 km altitude. At altitudes above 200 km, it is clear that the VLF whistler-mode wave dominates. Holzworth et al. (1999) showed that nearly all lightning strokes within a 2500 km radius of any point in the ionosphere produced upward-going whistler-mode plasma waves.

The study of the direct influence of lightning-generated waves on the magnetosphere is an active area of research, with many new models and data sets being generated even as this book is being written. Because of the transient nature of these lightning-generated waves and the high cost of high-resolution satellite-borne waveform measurements, we have only just begun to parse the direct relation between lightning-generated waves and the ambient plasma sphere and radiation belt particles in the magnetosphere.

2.6 RED SPRITES, ELVES, AND THE JETS

A variety of phenomena associated with lightning effects above the stratosphere were shown in Fig. 1.4. Together, these phenomena are referred to as TLEs (transient luminous events). Of these TLEs, red sprite events were imaged first and are typically detected over the positive CG flashes associated with extensive anvil clouds, whereas intense lightning strokes of either polarity can cause elves, which are rapidly expanding luminous rings at about 95 km altitude. The Great Plains of the United States are ideal for their detection due to both the geometry and the high level of activity in summer. From the foothills of the Rocky Mountains, one can look eastward over the horizon, which shields the sensitive cameras from direct lightning emissions of distant storms. The sprites are very bright, brighter than the aurora, but only last a few milliseconds. Winckler (1995 and references therein) was the first to record low-light-level TV ground photography of red sprites. They have been detected from airplanes (Sentman and Wescott, 1993), but the phenomenon really became accessible to intense scientific study after Lyons (1994) demonstrated an easy way to image sprites from the ground. The upper tips of sprites can reach above 90 km and the fine structure is remarkable. Figure 2.11 is a telescope's view of a 1 km × 1 km section showing many tendrils and bifurcations (Gerken et al., 2000).

Although no *in situ* electric field measurements have yet been made in an active sprite within the mesosphere, we can understand the red sprite phenomenon by looking at the electrostatic component of Eq. (2.2), which points downward in

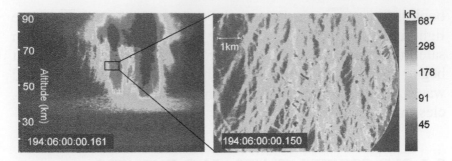

Fig. 2.11 Wide (left-hand panel) and telescopic (right-hand panel) field-of-view (FOV) false-colored images of a bright sprite event over Mexico. The narrow FOV with respect to the wide FOV is outlined by a black rectangle in the left-hand panel. This telescopic image demonstrates that, while the sprite appears to be amorphous in the wide FOV image, the lower portion of this sprite, in fact, consists of a volume of densely packed filaments. (For a color version of this figure, the reader is referred to the online version of this chapter.) Adapted from Gerken et al. (2000). Reproduced with permission of the American Geophysical Union.

the mesosphere following a positive CG stroke. The field in the mesosphere lasts for several milliseconds (see Fig. 2.2) until the local ambient conductivity has a chance to screen away the field. The discharge starts at the altitude where the neutral density has decreased sufficiently for a free electron to gain enough extra energy during the pulse and, while the electron is still between collisions, to cause impact ionization at the next collision. This may explain why positive CG lightning is more effective than the much more common negative lightning: during the positive CG lightning, the electric field in the mesosphere points downward, accelerating free electrons upward into the less dense air rather than down into the more dense air. In any event, the phenomenon proceeds upward and downward after initiation, much like a lightning streamer in the cloud prior to a lightning return stroke.

2.7 MAPPING OF MAGNETOSPHERIC DC ELECTRIC FIELDS TO THE STRATOSPHERE

Compared to the earth's radius of 6371 km, the altitude of the bottom of the ionosphere is very small (<100 km). Therefore, any large-scale magnetospheric electric field that exists in the ionospheric plasma and is caused, for instance, by planetary-scale motions of the plasma across field lines will have fringing fields that penetrate into the nearly neutral atmosphere down to the highly conducting earth's surface, depending on their scale size. This is analogous to a finite-sized parallel plate capacitor where fringing fields can easily be calculated in the steady state. The penetration, called mapping of the fields, will be modified by the conducting atmosphere, which affects both the amplitude and the allowed scale size of the penetrating electric field. This phenomenon was proven experimentally by Mozer and Serlin (1969) and Kelley et al. (1975). Following Mozer (1971), we consider the atmosphere to be a medium having a scalar conductivity that depends on altitude z:

$$\sigma(z) = \sigma_0 \exp(z/h).$$

Then, combining the following equations,

$$\nabla \times E = 0,$$
$$\nabla J = 0,$$
$$J = \sigma E,$$

and using

$$E = -\nabla V$$

yields a partial differential equation,

$$\frac{\partial^2 V}{\partial^2 x} + \frac{\partial^2 V}{\partial^2 y} + \frac{\partial^2 V}{\partial^2 z} + \frac{\partial V}{h \partial z} = 0.$$

If we impose a horizontal electric field of value E_2 at 100 km, the solution to this equation is of the form:

$$E_x(x,z) = E_2 \frac{1 - \exp(-z/h)}{1 - \exp(z_0/h)}.$$

The attenuation factor represented by the fraction shows how efficient the mapping is. In fact, the large-scale ionospheric and magnetospheric electric fields map to within a few scale heights of the surface. In the polar cap, auroral zone, and even at low latitudes under the solar quiet ionospheric dynamo, this downward mapping results in large-scale (100-1000s of km) horizontal electric fields at 10-30 km and above, which are nearly as large as the parent fields in the ionosphere (very little attenuation). Thus, the dc electric field in the stratosphere, far away from thunderstorms, will be dominated by the ionospheric/magnetospheric electric fields in the horizontal component, whereas the vertical electric field will be dominated by the global circuit return current (as discussed in Section 2.1).

REFERENCES

Bering III, E.A., Few, A.A., Benbrook, J.R., 1998. The global electric circuit. Phys. Today. 51 (10), 24–30.

Briggs, M.S., Connaughton, V., Wilson-Hodge, C., Preece, R.D., Fishman, G.J., Kippen, R.M., Bhat, P.N., Paciesas, W.S., Chaplin, V.L., Meegan, C.A., von Kienlin, A., Greiner, J., Dwyer, J.R., Smith, D.M., 2011. Electron-positron beams from terrestrial lightning observed with Fermi GBM. Geophys. Res. Lett. 38 (2), L02808. http://dx.doi.org/10.1029/2010GL046259.

Chalmers, J.A., 1967. Positive charges from the earth and the maintenance of the earth's fine-weather potential gradient. J. Atmos. Terr. Phys. 27, 307–310.

Connaughton, V., et al., 2013. Radio signals from electron beams in terrestrial gamma-ray flashes. Geophys. Res. Lett. 118 (5), 2313–2320. http://dx.doi.org/10.1029/2012JA018288.

Coulomb, C.A., 1795. Mém. de l'Acad., Paris. 616.

Elster, J., Geitel, H., 1899a. Phys. Z. 1, 245.

Elster, J., Geitel, H., 1899b. Über die Existenz elektrischer Ionen in der Atmosphäre (On the existence of electrical ions in the atmosphere). Terr. Mag. Atmos. Elect. 4, 213–234.

Gerken, E.A., Inan, U.S., Barrington-Leigh, C.P., 2000. Telescopic imaging of sprites. Geophys. Res. Lett. 27 (17), 2637–2640. http://dx.doi.org/10.1029/2000GL000035.

Gringel, W., Rosen, J.M., Hofmann, D.J., 1986. Electrical structure from 0 to 30 kilometers. In: The Earth's Environment: Studies in Geophysics. National Academy Press, Washington, DC, pp. 166–182.

Hale, L.C., 1984. Middle atmosphere electrical structure, dynamics and coupling. Adv. Space Res. 4 (4), 175–186.

Helliwell, R.A., 1965. Whistlers and Related Ionospheric Phenomena. Stanford University Press, Stanford, CA.

Hess, V.F., 1911. Phys. Z. 12, 998.

Holzworth, R.H., Winglee, R.M., Barnum, B.H., Li, Y., Kelley, M.C., 1999. Lightning whistler waves in the high-latitude magnetosphere. J. Geophys. Res. 104 (A8), 17369–17378. http://dx.doi.org/10.1029/1999JA900160.

Kelley, M.C., Tsurutani, B.T., Mozer, F.S., 1975. Properties of ELF electromagnetic waves in and above the Earth's ionosphere from plasma wave experiments on the OVI-17 and OGO-6 satellites. J. Geophys. Res. 80 (34), 4603–4611.

Kelley, M.C., Ding, J.G., Holzworth, R., 1990. Intense ionospheric electric and magnetic field pulses generated by lightning. Geophys. Res. Lett. 17 (12), 2221–2224.

Kelley, M.C., Vlasov, M.N., Foster, J.C., Coster, A.J., 2004. A quantitative explanation for the phenomenon known as storm-enhanced density. Geophys. Res. Lett. 31 (19), L19809. http://dx.doi.org/10.1029/2004GL020875.

Krehbiel, P., 1986. The electrical structure of thunderstorms. In: Krider, E.P., Robble, R.G. (Eds.), The Earth's Electrical Environment. National Academy Press, Washington, DC, pp. 90–113.

Krider, E.P., 1986. Physics of lightning. In: Krider, E.P., Roble, R.G. (Eds.), The Earth's Electrical Environment. National Academy Press, Washington, DC, pp. 30–40.

Le Monnier, L.-G., 1752. Observations sur l'Electricité de l'Air. Histoire de l'Académie royale des sciences, 233–243.

Lin, Y.T., Uman, M.A., Standler, R.B., 1980. Lightning return stroke models. J. Geophys. Res. 85 (C3), 1571–1583. http://dx.doi.org/10.1029/JC085iC03p01571.

Linss, W., 1887. Ueber einige die Wolken- und Luftelectricität betreffende Probleme. Meteorol. Z. 4, 345–362.

Lyons, W.A., 1994. Low-light video observations of frequent luminous structures in the stratosphere above thunderstorms. Mon. Weather Rev. 122, 1940–1946.

Mauchly, S.J., 1923. On the diurnal variation of the potential gradient of atmospheric electricity. Terr. Mag. Atmos. Elect. 28 (3), 61–81.

Moore, C.B., Vonnegut, B., 1977. The thundercloud. In: Golde, R.H. (Ed.), Lightning, vol. 1. Academic Press, New York, pp. 51–98 (Chapter 3).

Mozer, F.S., 1971. Balloon measurement of vertical and horizontal atmospheric electric fields. Pure Appl. Geophys. 84 (1), 32–45. http://dx.doi.org/10.1007/BF00875450.

Mozer, F.S., Serlin, R., 1969. Magnetospheric electric field measurements with balloons. J. Geophys. Res. 74 (19), 4739–4754.

Neher, H.V., 1961. Cosmic-ray knee in 1958. J. Geophys. Res. 66 (12), 4007–4012. http://dx.doi.org/10.1029/JZ066i012p04007.

Neher, H.V., 1967. Cosmic-ray particles that changed from 1954 to 1958 to 1965. J. Geophys. Res. 72 (5), 1527–1539. http://dx.doi.org/10.1029/JZ072i005p01527.

Rakov, V.A., Uman, M.A., 2003. Lightning: Physics and Effects. Cambridge University Press, Cambridge, MA.

Roble, R.G., Tzur, I., 1986. The global atmospheric–electrical circuit. In: Krider, E.P., Roble, R.G. (Eds.), The Earth's Electrical Environment. National Academy Press, Washington, DC, pp. 206–231.

Sentman, D.D., Wescott, E.M., 1993. Observations of upper atmospheric optical flashes recorded from an aircraft. Geophys. Res. Lett. 20 (24), 2857–2860. http://dx.doi.org/10.1029/93GL02998.

Stix, T.H., 1962. The Theory of Plasma Waves. McGraw-Hill, New York.

Stolzenburg, M., Marshall, T.C., 2008. Serial profiles of electrostatic potential in five New Mexico thunderstorms. J. Geophys. Res. 113 (D13), D13207. http://dx.doi.org/10.1029/2007JD009495.

Uman, M.A., 1969. Lightning. McGraw-Hill, New York.

Uman, M.A., 1987. The Lighting Discharge, International Geophysics Series, vol. 39. Academic Press, San Diego, CA.

Vlasov, M.N., Kelley, M.C., 2010. Crucial discrepancy in the balance between extreme ultraviolet solar radiation and ion densities given by the international reference ionosphere model. J. Geophys. Res. 115 (A8), A08317. http://dx.doi.org/10.1029/2009JA015103.

Volland, H., 1984. Atmospheric Electrodynamics, 11th ed. Springer-Verlag, Berlin/Heidelberg, Germany, pp. 4–28.

Whipple, F.J.W., Scrase, F.J., 1936. Point discharge in the electric field of the Earth. Geophys. Mem. (London) VIII (68), 1–20.

Wilson, C.T.R., 1900. On the comparative efficiency as condensation nuclei of positively and negatively charged ions. Philos. Trans. R. Soc. Lond. A. 193, 289–308. http://dx.doi.org/10.1098/rsta.1900.0009.

Wilson, C.T.R., 1920. Investigations on lightning discharges and on the electric field of thunderstorms. Philos. Trans. R. Soc. Lond. A. 221, 73–115.

Winckler, J.R., 1995. Further observations of cloud-ionosphere electrical discharges above thunderstorms. J. Geophys. Res. 100 (D7), 14335–14345. http://dx.doi.org/10.1029/95JD00082.

Woosley, J.D., Holzworth, R.H., 1987. Electrical potential measurements in the lower atmosphere. J. Geophys. Res. 92 (D3), 3127–3134. http://dx.doi.org/10.1029/JD092iD03p03127.

CHAPTER 3

Collisionless Hydrodynamic Generators in a Planetary Atmosphere

The Earth's Electric Field. http://dx.doi.org/10.1016/B978-0-12-397886-8.00003-X

3.1 INTRODUCTION

In this chapter, we study how atmospheric winds and waves in the 90-300-km height range create electric fields. The processes discussed here dominate the earth's electric field below 60° latitude. The phenomena discussed in Chapters 4 and 5 dominate above this latitude. In each case, the key diagnostic equation is that the divergence of current must vanish— basically Kirchhoff's Current Law from electrical circuit theory and conservation of charge from a physics viewpoint. In Section 3.2, we discuss the motion of the dense neutral upper atmosphere and how its interaction with the plasma creates electrical currents. For these currents to be divergence-free, electric fields must exist, and we provide some simplified models that predict the magnitude and direction of these fields. To proceed further, we need a better understanding of these atmospheric motions, which are better characterized by the terms "tides" and "internal waves." After this discussion, we examine how such waves can generate electric fields and other issues relating to the earth's electric field.

3.2 COLLISIONAL HYDRODYNAMIC GENERATION OF ELECTRIC FIELDS

3.2.1 Upper Thermospheric Generators (The F-Region Ionosphere)

Figure 3.1 shows typical profiles of atmospheric temperatures and ionospheric plasma density for earth. However, the earth's thermosphere is so tenuous that it has no thermal memory and hence displays a huge daily variation in neutral temperatures, T_n. On a given day, T_n will peak to above 1000 K at around 14:00 LT at a given location and fall to well under 1000 K near midnight, creating a huge pressure gradient pointing from the nightside to the dayside. Thus, it is not surprising that high winds occur across the planet's upper atmosphere. It is interesting that Mars' atmosphere is very tenuous, even at the surface, and also exhibits large daily variations in wind speed from day

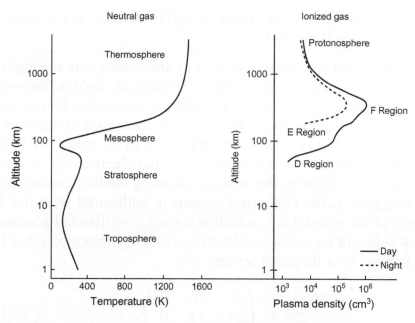

Fig. 3.1 Typical profiles of neutral atmospheric temperature and ionospheric plasma density with the various layers designated. After Kelley (2009). Reproduced with permission of Elsevier.

to night (Chamberlain and Hunten, 1987). Thermospheric wind speeds for earth easily reach 30% of the speed of sound on a daily basis whereas, at the surface, the winds are much more localized (i.e., not global) and are much less than the speed of sound. The term "thermosphere" comes from the rapid rise in the neutral temperature due to the absorption of UV and EUV from the sun at high altitudes. This absorption protects the planet from deadly rays, as does the ozone layer. Note the increase in temperature near 50 km due to this ozone effect.

These emissions from the sun also ionize the upper atmosphere and thereby create the ionosphere during the day, along with increasing the temperature. The wind blows away from the corresponding high pressure area, antiparallel to the pressure gradient. This is quite different from the lower atmosphere where winds, to first order, are perpendicular to the pressure gradient. We discuss the reason for this difference below. Since the heating is diurnal, we thus have a diurnal "internal tide" in the upper atmosphere (roughly defined as the height range of 180-1000 km).

The easiest electrical generator to understand is in the night-time equatorial ionosphere where the magnetic field is horizontal and points toward the north. (In the discussion below, the y-axis is parallel to the magnetic field, the x-axis is eastward, and the z-axis is upward. The declination of the magnetic field is taken to be zero.) Imagine now that the electric field is zero and a wind is blowing through a conducting medium containing a magnetic field. This wind system is collocated with the F layer of the ionosphere, which is a weakly collisional plasma. In this height zone, the conductivity is highly anisotropic but is described by a diagonal tensor,

$$\sigma = \begin{pmatrix} \sigma_P & 0 & 0 \\ 0 & \sigma_0 & 0 \\ 0 & 0 & \sigma_P \end{pmatrix}, \tag{3.1}$$

where σ_P is the so-called Pedersen conductivity, which is controlled by the ions

$$\sigma_P = \frac{ne^2 v_{in}}{M\Omega_i^2},$$

and σ_0 is the parallel conductivity,

$$\sigma_0 = \frac{ne^2}{mv_{en}},$$

which is controlled by the electrons. Note that, near the equator, we take **B** parallel to the y-axis, so Eq. (3.1) differs from Eq. (1.9). In these expressions, the Ω's are gyrofrequencies, M and m are the mean ion and electron masses, and v is the collision frequency with neutrals of the ions (i) and electrons (e). The ion and electron densities are almost equal and are denoted by "n." (To read more about the ionospheric conductivity tensor, see Kelley, 1989, 2009.)

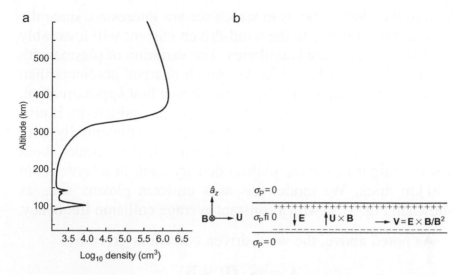

Fig. 3.2 (a) Typical post-sunset plasma density profile at the equator and (b) a slab geometry representing the profile characteristics near 350 km. After Kelley (2009). Reproduced with permission of Elsevier.

Now consider the equatorial ionospheric F layer, which has a density profile similar to that shown in Fig. 3.2a. The Pedersen conductivity will also vary drastically with altitude and, in a simple model, looks like a slab of high conductivity, as in Fig. 3.2b.

So how does the wind drive a current? First, note that σ_0 is very high and, for any reasonable currents along the magnetic field, the corresponding electric field must be small. Thus, the component of the wind parallel to $\mathbf{B}$ only pushes the plasma along the field lines and no electric field of any significance is generated. So, we are interested in the wind component across the magnetic field, $\mathbf{U}_\perp$. Consider a uniform plasma with a constant wind and no electric field in the earth-fixed frame, that is, $\mathbf{E} = 0$ in Eq. (1.4). If we move into the frame moving with velocity $\mathbf{U}_\perp$, then $\mathbf{E}' = \mathbf{U}_\perp \times \mathbf{B}$. In this frame, $\mathbf{J}' = \sigma_P \times \mathbf{E}' = \sigma_P(\mathbf{U}_\perp \times \mathbf{B})$, so there is a current in the primed frame. But using Eq. (1.15), $\mathbf{B} = \mathbf{B}'$ and there must be the same current in both frames. So, in the earth frame, it must also be the case that $\mathbf{J} = \sigma_P(\mathbf{U}_\perp \times \mathbf{B})$.

So far the notion of an infinite, uniform system is only a thought exercise. Any realistic situation has boundaries that lead to the electric fields in which we are interested, since the charge transport due to the wind-driven current will invariably be deposited near the boundaries. The variation of plasma with altitude shown in Fig. 3.2a has much sharper gradients than occur in either horizontal direction, so as a first approximation, we consider these variations. The Pedersen conductivity is proportional to nv_{in}. The second term varies exponentially with altitude and it is straightforward to show that σ_P peaks about a scale height below the plasma density peak in a layer about 100 km thick. We model this as a uniform plasma slab, as shown in Fig. 3.2a, with a constant average collision frequency.

As noted above, the wind-driven current is given by:

$$\mathbf{J} = \sigma_P \cdot (\mathbf{U} \times \mathbf{B}).$$

The cross product is orthogonal to $\mathbf{B}$ and taking $\mathbf{U}$ eastward gives a vertically upward current. Now, given the profiles in Figs. 3.1 and 3.2a, clearly $\nabla \cdot \mathbf{J} \neq 0$ and an electric field must exist so that:

$$\nabla \cdot [\sigma \cdot (\mathbf{E} + \mathbf{U} \times \mathbf{B})] = 0. \tag{3.2}$$

For the moment, if we ignore any leakage of current along the magnetic field, a simpler result pertains, namely:

$$\mathbf{E} + \mathbf{U} \times \mathbf{B} = 0$$

or

$$\mathbf{E} = -\mathbf{U} \times \mathbf{B}. \tag{3.3}$$

which is vertically downward. Using Eq. (1.12),

$$\mathbf{V}_\perp = \frac{\mathbf{E} \times \mathbf{B}}{B^2}.$$

Then substituting $\mathbf{E} = -\mathbf{U} \times \mathbf{B}$ gives

$$\mathbf{V}_\perp = -\frac{(\mathbf{U} \times \mathbf{B}) \times \mathbf{B}}{B^2} = \mathbf{U}_\perp.$$

The ions, electrons, and neutrals move at the same speed. Since $U_\perp$ is eastward, the plasma velocity will also be eastward.

To include magnetic field effects, we need to take a three-dimensional viewpoint and include possible leakage of $J_\perp$ via $J_\parallel$ to the ends of the field lines. Since the field lines are in the y-direction, the important equation is:

$$(\nabla_\perp \cdot \mathbf{J}) + \frac{dJ_y}{dy} = 0,$$

$$J_y = - \int (\nabla \cdot \mathbf{J}_\perp) dy.$$

(3.4)

Before considering this equation, recall how electric fields map in magnetized plasmas. For a large parallel conductivity, the potential across any two magnetic field lines is unchanged and maps down to both E regions. To understand the diurnal variation, however, we need to consider the role of the "end" plates in the E layer.

At night, the E-region conductivity is low and the high zonal winds at altitudes of several hundred kilometers determine the vertical electric field and hence the horizontal plasma flow. For $\Sigma_P^F \gg \Sigma_P^E$, the electric field will be almost equal to uB, where u is the eastward wind speed. The eastward plasma velocity thus nearly matches the neutral wind speed. During the day, however, the F-region upper thermospheric dynamo loses control of the electrodynamics and the resulting electric fields are determined by winds in the E region. A simple model for the diurnal variation is discussed next.

An analogous electric circuit is shown schematically in Fig. 3.3. The three batteries correspond to the two (north and south) E-region dynamos (see Section 3.2.2) and the F-region dynamo. Each battery has a finite internal resistance. The voltage measured by the meter corresponds to the electric field times the distance between the two magnetic field lines (the conducting wires). The battery with the lowest internal resistance determines the voltage and hence the electric field.

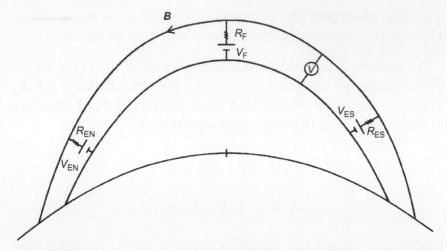

Fig. 3.3 Electric circuit analogy to the voltage sources in the equatorial F region. The off-equatorial E-region wind dynamo competes with the F-region dynamo at the equator to determine the voltage differences between magnetic field lines. After Kelley (2009). Reproduced with permission of Elsevier.

In the F layer during the nighttime, $R_F < R_E$ and the F region dominates. During the day, $R_E < R_F$ and E-region wind sources determine the electric field and the F layer acts as a load.

3.2.2 The Global E-Region Electric Field Generator

In the E region of the ionosphere, the conductivity tensor is not diagonal and the Hall terms, σ_H, must also be considered. (Remember that, when comparing with Eq. (1.19), we orient the magnetic field in the y-direction at the equator.)

$$\sigma = \begin{pmatrix} \sigma_P & 0 & \sigma_H \\ 0 & \sigma_0 & 0 \\ -\sigma_H & 0 & \sigma_P \end{pmatrix}.$$

Furthermore, the electric field cannot be taken as being entirely self-generated by local wind fields, as we assumed for the night-time F-layer dynamo. This can be understood as follows. The entire dayside ionosphere is a good electrical conductor in which currents are driven by lower thermospheric tides. The tidal E-region wind field $\mathbf{U}(\mathbf{r}, t)$ is global in nature and will drive a global current system given by $\mathbf{J_w} = \sigma(\mathbf{r}, t) \cdot [\mathbf{U}(\mathbf{r}, t) \times \mathbf{B}]$. Now, since both $\sigma(\mathbf{r}, t)$ and $\mathbf{U}(\mathbf{r}, t)$ depend on $\mathbf{r}$, the current $\mathbf{J_w}$ is not, in general, divergence-free. Thus, an electric field $\mathbf{E}(\mathbf{r}, t)$ must build up such that the divergence of the total current is zero and

$$\nabla \cdot [\sigma(\mathbf{E}(\mathbf{r}, t) + \mathbf{U}(\mathbf{r}, t) \times \mathbf{B})] = 0. \tag{3.5}$$

The resulting $\mathbf{E}(\mathbf{r}, t)$ is as rich and complex as the driving wind field and the conductivity pattern that produces it. The latter has a primarily diurnal variation over the earth's surface, whereas the dominant tidal winds change from diurnal to semidiurnal, depending on latitude. As a first approximation, we might then expect a diurnal electric field pattern at low latitudes and a mixture of diurnal and semidiurnal electric fields at higher latitudes. This crude analysis actually describes the situation fairly well (see Fig. 3.4, which shows E fields at four latitudes).

The diurnal variations of the electric field components measured over Peru at the Jicamarca Observatory are shown in Fig. 3.4a and are dominated by diurnal effects (see Chapter 7 for a discussion of how radar observatories can determine the electric field). Both the plasma drift and the electric field components are indicated in the plot scaling. The global vertical magnetic perturbation pattern measured by ground-based magnetometers is shown in Fig. 3.5 and gives further evidence for this simple diurnal picture of equatorial electromagnetics. High magnetic fields occur only during the day because of the high conductivity, but the electric field is present both day and night. Such data have been used to construct the pattern of electrical currents in the entire ionosphere that is referred to as the Sq current system. (S stands for solar and q for quiet in this notation.) Ground magnetometers respond primarily to horizontal Hall currents since the magnetic field due to field-aligned currents is canceled out by the Pedersen currents that link the field lines horizontally through the ionosphere. Since the electric field must be nearly the same at both ends of a magnetic field line, any differences between the two hemispheres create currents that flow between the hemispheres to maintain

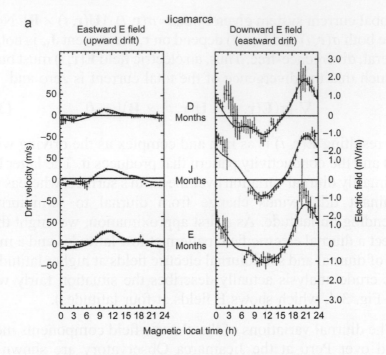

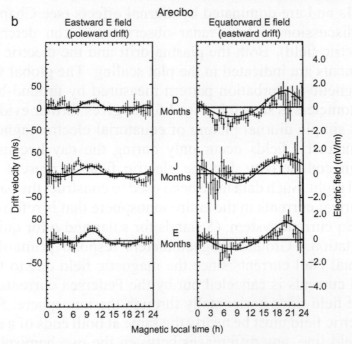

Fig. 3.4 Seasonally averaged quiet-day drifts and electric fields (points with error bars) and model drifts (solid lines) perpendicular to the geomagnetic field at 300 km for (a) Jicamarca, Peru, (b) Arecibo, Puerto Rico,

Continued

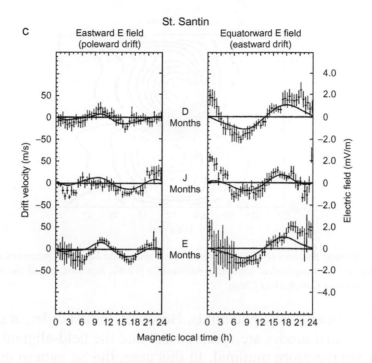

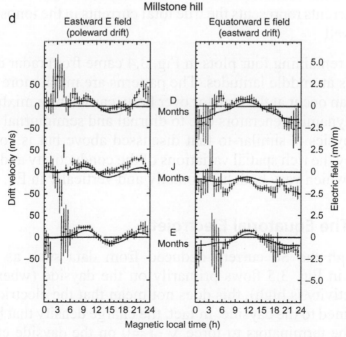

Fig. 3.4, cont'd (c) St. Santin, France, and (d) Millstone Hill, MA. D refers to Northern Hemisphere winter, J to Northern Hemisphere summer, and E to equinox. After Richmond et al. (1980). Reproduced with permission of the American Geophysical Union.

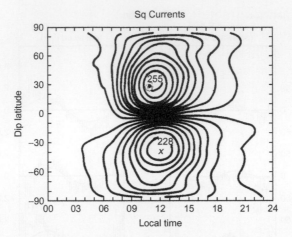

Fig. 3.5 Average contours of vertical magnetic field due to the Sq system measured during the International Geophysical Year. After Matsushita (1969). Reproduced with permission of the American Geophysical Union.

the field lines as equipotentials. However, to first order, at equinox the tidal modes are symmetrical and the field-aligned currents are therefore minimal. In this case, the Sq pattern due to Hall currents represents the true total currents in the ionosphere quite well.

The remaining four plots in Fig. 3.4 came from radar observatories at middle latitudes. The patterns are much more complex than occur at the equator over Peru, caused by a mixture of hydrodynamic generators due to diurnal and semidiurnal tides. The physics is similar to that discussed above but is complicated by the rich spatial variations of the conductivity and wind fields, which vary both horizontally and vertically in Eq. (3.5).

3.2.3 The Equatorial Electrojet

Although the Sq current deduced from data such as those shown in Fig. 3.5 flows primarily on the dayside (where the conductivity is high), this does not mean that the electric field is confined to the dayside. In fact, the charge density that builds up at the terminators to force $\nabla \cdot \mathbf{J} = 0$ on the dayside creates the nighttime zonal electric field pattern observed at Jicamarca.

We consider the effect of the eastward zonal electric field at the equator. The dense contour lines at the equator in Fig. 3.5 are due to a current component called the equatorial electrojet.

Although small, the zonal electric field component in Fig. 3.4 has important consequences. Of course, it will drive a small Pedersen current eastward along the dayside equator. More importantly, a vertically downward Hall current will also flow due to this electric field. Now from $\nabla \cdot \mathbf{E} = 0$, we can deduce that

$$\frac{\partial E_x}{\partial z} = \frac{\partial E_z}{\partial x}.$$

This means that the variation of the zonal component with altitude can be estimated by the ratio

$$\delta E_x = \delta E_z \left(\frac{\delta z}{\delta x} \right).$$

The scale size (δx) of the horizontal conductivity pattern is 100 times that of the vertical (δz) variations of conductivity, whereas both experiments and theory indicate E_z is at most 10-20 times E_x. Taken together, this means that the zonal electric field can change only slightly in the E region, and we thus will assume that conductivity gradients rather than variations in E_x dominate the divergence of the vertical Hall current. To investigate this divergence, the altitude variations of typical noontime equatorial electrojet conductivities and densities are plotted in Fig. 3.6 (Forbes and Lindzen, 1976). Notice that the Hall conductivity dominates below 120 km and, indeed, is highly altitude dependent, even though the plasma density itself is nearly uniform with height.

As a first approximation to the physics of the E-region dynamo, we again consider a slab conductivity geometry, such as that illustrated in Fig. 3.7, subject to a constant zonal electric field. The Hall current cannot flow across the boundary, and charge layers must build up, generating an upward-directed electric field. In a steady state in this slab model, no vertical

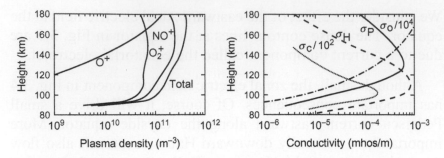

Fig. 3.6 Vertical profiles of daytime composition and plasma density (left) and conductivities (right) for average solar conditions. After Forbes and Lindzen (1976). Reproduced with permission of Pergamon Press.

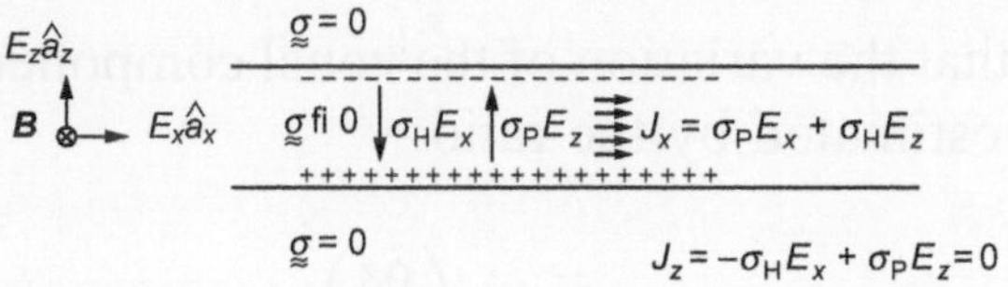

Fig. 3.7 The equatorial electrojet in a slab geometry. After Kelley (2009). Reproduced with permission of Elsevier.

current may flow and the vertical Pedersen current must exactly cancel the Hall current. This implies that

$$\sigma_H E_x = \sigma_P E_z$$

and hence that

$$E_z = \left(\frac{\sigma_H}{\sigma_P}\right) E_x. \tag{3.6}$$

Since $\sigma_H > \sigma_P$, the vertical electric field component considerably exceeds the zonal electric field component. In addition, $E_z(z)$ has the same z dependence as the function $\sigma_P(z)/\sigma_H(z)$. The zonal current is now given by

$$J_x = \sigma_H E_z + \sigma_P E_x, \tag{3.7a}$$

$$J_x = \left[\left(\frac{\sigma_H}{\sigma_P}\right)^2 + 1\right] \sigma_P E_x = \sigma_c E_x, \tag{3.7b}$$

where σ_c is the so-called Cowling conductivity. Notice that the local neutral wind does not enter this calculation yet at all; the electrojet is set up by the global tidal winds that create the diurnal zonal electric field component measured at the equator. In a more complete theory, complications due to the zonal neutral wind may be included. Note that the meridional wind component does not enter at the magnetic equator, since for that component, the cross product with the magnetic field vanishes.

In effect, Eq. (3.7b) shows that the zonal conductivity is enhanced by the large factor $1 + \sigma_H^2/\sigma_P^2$, the Cowling conductivity factor, which leads to the intense current jet at the magnetic equator. This can be seen in Fig. 3.5, in which the magnetic field contours become very close together at the magnetic equator. This channel of electrical current is termed the equatorial electrojet. The Cowling conductivity is also plotted in Fig. 3.6 (divided by 100) and displays a peak at 102 km with a half-width of 8 km.

Before leaving this phenomenon, we will include not only an eastward electric field but a zonal wind. Why? For reasons not completely understood, winds in this height range (100-110 km) are actually quite large. Hundreds of examples of E-region winds determined from the motion of trimethyl aluminum (TMA) trails are superposed in Fig. 3.8 (upper panel). Winds are often over 100 m/s.

Suppose only a wind exists without any electric field. Then,

$$\mathbf{J}_U = \begin{pmatrix} \sigma_P & 0 & \sigma_H \\ 0 & \sigma_0 & 0 \\ \sigma_H & 0 & \sigma_P \end{pmatrix} \begin{pmatrix} 0 \\ 0 \\ uB \end{pmatrix} = \begin{pmatrix} \sigma_H uB \\ 0 \\ +\sigma_P uB \end{pmatrix}.$$

In this height range, $\sigma_H \cong ne/B$, so the eastward current is $+ncu$. This equation is easy to interpret; the wind simply carries the ions along and the electrons stay still. It is interesting to study this situation in the wind reference frame. There,

$$\mathbf{E}' = 0 + \mathbf{U} \times \mathbf{B}.$$

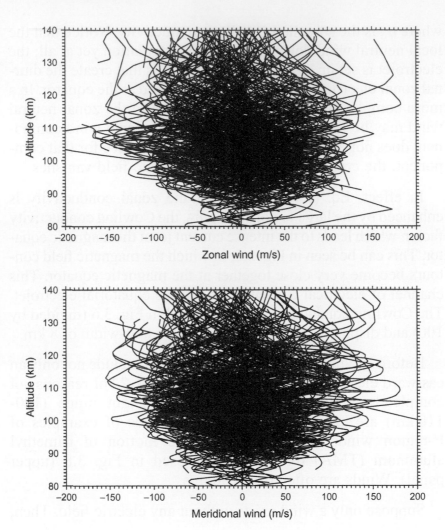

Fig. 3.8 (Upper panel) Superposition of numerous wind profiles from TMA trails. (Lower panel) Distortion of a visible meteor train by an inertio-gravity wave. After Larsen (2002). Reproduced with permission of the American Geophysical Union.

In this reference, the ions are at rest and the electrons $(\mathbf{E}' \times \mathbf{B}/B^2)$ drift. Since $\mathbf{E}'$ is up, $\mathbf{E}' \times \mathbf{B}/B^2$ is west with magnitude $\mathbf{u}$. Electrons moving west at velocity $\mathbf{U}$ carry an eastward current with magnitude neu. Note that $\mathbf{J}' = \mathbf{J}$, as argued earlier in Section 1.4 in Chapter 1.

There have been many rocket measurements for the magnetic field due to the electrojet current. Some of the data are

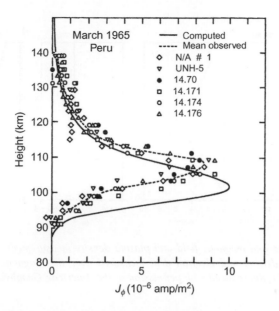

Fig. 3.9 Observed and computed eastward current density profiles near noon at the dip equator off the coast of Peru in March 1965, normalized to a magnetic field perturbation of 100 nT at Huancayo. Measured profiles are from Shuman (1970) (flight N/A #1), Maynard (1967) (flight UNH-5), and Davis et al. (1967) (flights 14.170, 14.171, 14.174, and 14.176). The theoretical profile is from Richmond's theory (Richmond, 1973a). After Richmond (1973b). Reproduced with permission of Pergamon Press.

reproduced here in Fig. 3.9. The plot displays the altitude variation of the jet over Peru with each profile normalized to a 100-nT variation of the magnetic field measured on the ground at Huancayo, Peru; that is, the actual perturbed magnetic field at Huancayo during each rocket flight was used to scale all data to a common ground perturbation of 100 nT.

Of more importance to the present text is the generation of large vertical electric fields in the electrojet, an example of which is presented in Fig. 3.10 along with the electron density and horizontal current (Pfaff et al., 1997). As discussed in Chapter 6, these high fields create electron drifts that are unstable to the generation of plasma waves. Peak vertical electric fields in the range of 10-20 mV/m have been predicted theoretically and would create electron drift speeds of 350-700 m/s. This exceeds the acoustic speed C_s, which is about 360 m/s,

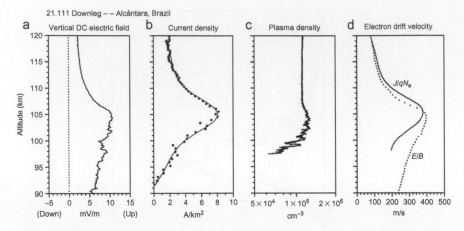

Fig. 3.10 Electric and magnetic field and plasma density measurements along with the derived current density and electron drift measured over the magnetic equator. After Pfaff et al. (1997). Reproduced with permission of the American Geophysical Union.

and leads to the generation of intense plasma waves via the two-stream instability. These rocket data, along with many years of radar observations, show that the vertical electric field reaches values at least as large as that given by $E_z = C_s B \simeq 9$ mV/m at the very threshold instability, and it is very likely that the field reaches values up to twice as large. This is the largest electric field found in the ionosphere below subauroral zone latitudes.

The measurements in panels (a-c) have been combined in the right-hand panel to show the differential (net current) velocity as well as the $\mathbf{E} \times \mathbf{B}$ drift speed. At high altitudes the agreement is quite good, but below 104 km there is a systematic difference. This suggests that a high neutral wind velocity may have been present, a result that is in agreement with TMA measurements 2 days later (Larsen and Odom, 1997) and with Fig. 3.8.

3.3 THE HYDROMAGNETIC GENERATOR SOURCE: THE RESTLESS ATMOSPHERE

Figure 3.11 shows a spectrum of horizontal neutral winds measured in a 2-year period over Kauai. Very clear oscillations are seen at what are called planetary wave periods 2-5 and 16-day

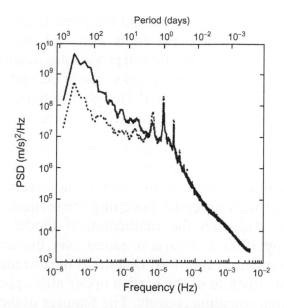

Fig. 3.11 The spectrum of mesospheric waves determined using 3 years of data. After Fritts and Isler (1994). Reproduced with permission of the American Meteorological Society.

waves are seen. Also seen are tides at 24 and 12 h. Then a region of power law exists, representing what are called internal buoyancy or gravity waves. The latter are somewhat randomly generated by cold fronts, mountain ranges, etc.

The equations of motion describing wave modes on a rotating planet are actually quite complicated. Many books have been written, including a classic by Chapman and Lindzen (1970) on atmospheric tides. In very brief form, we discuss this below.

Tidal theory is quite complex. The equations of the neutral atmosphere must be solved on a rotating spherical shell subject to the earth's gravitational field. Considerable insight is obtained by studying the free oscillations of the atmosphere, that is, the normal modes of the system. This is accomplished by reducing the set of equations to one second-order partial differential equation, which is often written in terms of the divergence of the wind field. The resulting equation is separable in terms of functions of latitude (θ), longitude (ϕ), altitude (z), and time (t). The longitude/time dependence is of the form $\exp[i(s\phi + ft)]$, where

s must be an integer. For $s=0$, the temporal behavior does not propagate with respect to the earth. For $s=1$, the disturbance has one oscillation in longitude and propagates westward following the sun; this is the diurnal tide. The θ dependence can be expressed in terms of so-called Hough functions, which may be related to the spherical harmonic functions.

The height dependence of these normal modes is a crucial factor in tidal theory, since when forcing (e.g., solar heating) is included in the equations, the only modes excited are those that have a vertical structure, matching the vertical structure of the forcing. Although the mathematical models were well developed by 1900, controversy raged over the actual nature of the atmospheric tides until definitive measurements of the temperature structure of the earth's upper atmosphere became available from sounding rockets. The features of these temperature profiles, which are of most relevance to tidal theory, are due to the absorption of sunlight by ozone in the stratosphere and by water vapor in the troposphere.

When both the θ and z dependences of forcing functions and the normal modes are taken into account, many of the dominant features of the tidal oscillations can be explained. For our purposes, we summarize these as follows:

1. Tidal oscillations propagate upward and the associated wind speed amplitude grows as they do so. (This amplification is due to the decreasing density of the atmosphere and is a consequence of energy conservation. This important feature of vertical wave propagation is discussed in detail in Chapter 5, where gravity waves are studied.)

2. Diurnal tides can propagate vertically only below 30° latitude. At higher latitudes, they remain trapped in the stratosphere.

3. With the decreasing importance of the diurnal tide, the semidiurnal tide becomes dominant at latitudes higher than 30°.

Armed with this modest understanding of tidal theory, we may now investigate some aspects of the E-region dynamo.

Before leaving this tidal discussion, we must go beyond propagating tides generated by the lower atmosphere. In the thermosphere, there is so little thermal motion that the wind simply blows across the earth from high pressure to low pressure. This is seen as a 24-h tide on the rotating planet. This is either an *in situ* tide or a diurnal wind, take your choice. This global tide is present more or less everywhere and competes with the propagating 24-h tide. The ionosphere is not passive in this *in situ* tide. During the day, the plasma content is stuck onto magnetic field lines since no electric field can be generated by the wind blowing on the plasma. Why? During the day, the F-region dynamo is shorted out by the dense E region. Every ion hit by a neutral atom simply moves one gyroradius and waits to be hit again. This is called ion drag. The force due to a current is equal to $\mathbf{J} \times \mathbf{B}$, so

$$\mathbf{F} = \mathbf{J} \times \mathbf{B} = \sigma(\mathbf{U} \times \mathbf{B}) \times \mathbf{B},$$

$$\mathbf{F} = \frac{ne^2 v_{\text{in}}}{M\Omega_i^2} \left(-\mathbf{U}_\perp B^2 \right),$$

where the σ_P approximation used is good for the F region. Thus,

$$\mathbf{F} = -n_e m_i v_{\text{in}} \mathbf{U}_\perp.$$

Clearly, this is a frictional drag effect. At night there is no E region, $\mathbf{E} = -\mathbf{U} \times \mathbf{B}$, and there is almost no ion drag since the plasma and neutrals move together.

For strong ion drag, the effect of the earth's rotation is somewhat suppressed. For example, the Navier-Stokes equation is:

$$\frac{d\mathbf{V}}{dt} = \frac{-\nabla p}{\rho} + \frac{2(\mathbf{\Omega} \times \mathbf{U})}{\rho} - \frac{nm_i v_i \mathbf{U}_\perp}{\rho}.$$

For no acceleration,

$$\nabla p = 2(\mathbf{\Omega} \times \mathbf{U}) - NM_i v_m \mathbf{U}_\perp.$$

If $NM_i v_m > 2 \Omega$, the second term dominates. For 300 km in daytime,

$$\frac{NM_i v_{in}}{\rho} \approx \frac{Ne}{Nn} v_m \approx \frac{10^{12}}{10^{16}} (1) = 10^{-4} s^{-1}.$$

This is larger than the Coriolis frequency, $2 f \sin\theta$, where θ is the latitude. The upshot is that the wind simply flows from high to low pressure with no rotational effect. This explains why thermospheric winds are parallel to $-\nabla p$, unlike in the troposphere where the wind is primarily perpendicular to ∇p, and that ionospheric drag is the cause.

At frequencies that are higher than the tides, the upper atmosphere is continuously bombarded with gravity waves from a number of sources, including tropospheric weather fronts, tornadoes and thunderstorms, impulsive auroral zone momentum injection and heating events, and even earthquakes and volcanic eruptions. The famous monograph entitled *The Upper Atmosphere in Motion* by Hines (1974) is an excellent annotated collection of gravity wave studies published by Hines and coworkers over about a 10-year period. The reader is referred to that work for details about gravity waves and tidal oscillations as well as to the excellent review of tidal theory by Chapman and Lindzen (1970) mentioned earlier. Here our approach is much more modest in scope, aiming at physical intuition rather than detailed analysis.

We study gravity waves first, in effect finding the normal modes of a flat nonrotating inviscid atmosphere. These results will be valid as long as the periods do not approach the tidal range, and the wavelengths are not long enough that the curvature of the earth matters. We assume an isothermal, inviscid atmosphere initially in hydrostatic equilibrium, so that if ρ_0 and p_0 are the zero-order mass density and pressure, the relation

$$\rho_0 \mathbf{g} = -\nabla p_0$$

applies. In addition, it can be shown that ρ_0 and p_0, which vary only in the vertical direction in this model, are of the form

$$\rho_0, \quad p_0 \propto e^{-z/H},$$

where H is the scale height of the atmosphere, that is, $1/H = -(1/\rho_0)(d\rho_0/dz)$. Here, we again choose our coordinates using the meteorological convention and take x eastward, y northward, and z vertically upward. We assume there are no neutral winds in the unperturbed atmosphere. The equations governing the behavior of the atmosphere are the mass continuity equation, the equation of motion, and the adiabatic condition (see Yeh and Liu, 1974). In the equation of motion, only terms due to gravity, pressure gradients, and inertia are retained.

Now consider atmospheric oscillations in the presence of gravity. We assume that there are small perturbations in the mass density, pressure, and wind velocity denoted by $\delta\rho$, δp, and $\mathbf{U} = (u, v, w)$. Without the Coriolis or viscous forces, there is no coupling between oscillations in the y-z plane and those in the x direction, so we can ignore the x component of velocity, making the problem two dimensional. For meridional propagation, we define a column vector, $\mathbf{F}$, by

$$\mathbf{F} = \begin{vmatrix} \delta\rho/\rho_0 \\ \delta p/p_0 \\ v \\ w \end{vmatrix}$$

and assume that atmospheric perturbations can be described by plane waves of the form

$$F \propto e^{i\left(\omega t - k_y y - k_z z\right)}. \tag{3.8}$$

Substituting $\rho = \rho_0 + \delta\rho$, $p = p_0 + \delta p$, $\mathbf{U} = (0, v, w)$ into the equations describing the atmosphere and retaining terms up to first order in $\delta\rho$, δp, and $\mathbf{U}$ give the linearized forms of the mass continuity, motion, and adiabatic state equations, that is,

$$\frac{\partial(\delta\rho)}{\partial t} + \mathbf{U}\cdot\nabla\rho_0 + \rho_0\nabla\cdot\mathbf{U} = 0, \tag{3.9a}$$

$$\frac{\rho_0\partial v}{\partial t} + \frac{\partial(\delta p)}{\partial y} = 0, \tag{3.9b}$$

$$\frac{\rho_0\partial w}{\partial t} + \frac{\partial(\delta p)}{\partial z} + \delta\rho g = 0, \tag{3.9c}$$

$$\frac{\partial(\delta p)}{\partial t} + \mathbf{U}\cdot\nabla p_0 - C_0^2\frac{\partial(\delta\rho)}{\partial t} - C_0^2\mathbf{U}\cdot\nabla\rho_0 = 0. \tag{3.9d}$$

We have taken the atmosphere to be isothermal with temperature T. In Eq. (3.9d), C_0 is the speed of sound, given by

$$C_0^2 = \frac{\gamma p_0}{\rho_0} = \gamma g H,$$

where γ is the ratio of specific heats at constant pressure and constant volume and $H = kT/Mg$ is the scale height. Viscosity has been ignored. Using Eq. (6.1) and the condition for hydrostatic equilibrium, Eq. (6.2) can be rewritten as a matrix equation:

$$\begin{vmatrix} i\omega & 0 & -ik_y & -1/H - ik_z \\ 0 & -ik_y C_0^2/\gamma & i\omega & 0 \\ g & -C_0^2(1/H + ik_z)/\gamma & 0 & i\omega \\ -i\omega C_0^2 & i\omega C_0^2/\gamma & 0 & (\gamma - 1)g \end{vmatrix} \cdot \mathbf{F} = 0. \tag{3.9e}$$

In deriving Eq. (3.9c), we used the fact that $\delta p \propto p_0 \exp(i\omega t - ik_y y - ik_z z)$. This leads to $\partial(\delta p)/\partial z = \delta p[(1/p_0)(dp_0/dz) - ik_z]$ and eventually to the corresponding entry in row 3. Setting the determinant of the 4×4 matrix equal to zero yields the dispersion relation for linear modes of a nonrotating neutral atmosphere on a flat earth,

$$\omega^4 - \omega^2 C_0^2\left(k_y^2 + k_z^2\right) + (\gamma - 1)g^2 k_y^2 + i\gamma g\omega^2 k_z = 0. \tag{3.10}$$

A variety of possible wave modes are buried in this dispersion relation. Suppose we take the limit that $g = 0$. Then Eq. (3.10) reduces to

$$\omega^2 = C_0^2 \left(k_y^2 + k_z^2 \right),$$

which is the dispersion relation for sound waves propagating without attenuation, growth (pure real ω and $\mathbf{k}$), or dispersion ($\omega/k =$ constant).

We now turn to the gravity wave case. If there are no sources of energy or dissipation (viscosity was ignored), waves will not grow or decay in time at a fixed point in space, so we can assume ω is real. If we are including gravity, however, it can be shown that there are no solutions of Eq. (3.10) with both k_y and k_z purely real. Anticipating the final result, let us assume k_y is purely real and investigate k_z. This corresponds to a wave propagating in an unattenuated fashion with a component in the horizontal direction. Then we can write Eq. (3.10) as

$$\omega^4 - \omega^2 C_0^2 k_y^2 + (\gamma - 1) g^2 k_y^2 = -i\gamma g \omega^2 k_z + \omega^2 C_0^2 k_z^2, \quad (3.11)$$

where the left-hand side is purely real. Now if we let k_z be a complex number,

$$k_z = k_z' + i k_z'',$$

it is straightforward to show that the right-hand side of Eq. (3.11) is purely real if and only if

$$k_z'' = (1/2H).$$

Dropping the superscript (prime) notation, we now can see that the solutions for the quantities in the column vector $\mathbf{F}$ are of the form

$$e^{i\left(\omega t - k_y y - k_z z\right)} e^{z/2H}. \quad (3.12)$$

In Eq. (3.12), ω, k_y, and k_z are real.

TMA Trail
June 11, 1978
0634 UT
(Wallops Island, VA)

Fig. 3.12 A TMA trail deployed from Wallops Island, Virginia on June 11, 1978 at 06:34 UT. The trail was photographed from the NASA C54 airplane. After Kelley (2009). Reproduced with permission of Elsevier.

Atmospheric waves that propagate in the manner described by Eq. (3.11) are termed internal gravity waves. Some of the complexity of the wind patterns that arise in the 90-120-km height range due to such waves can be gauged from the photograph of a TMA vapor trail deployed by a sounding rocket, shown in Fig. 3.12. This photograph yields only one perspective on the distortion of the trail by the ambient winds but numerous reversals and shears are evident. Figure 3.8a shows a superposition of many wind observations using the TMA technique. The winds are very large and very structured (Larsen, 2002).

From Eq. (3.12), the theoretical prediction is that the wave amplitude should grow as it propagates upward (positive z). The physical explanation for this somewhat bizarre result is that to conserve the wave perturbation energy (e.g., terms of the form $\rho_0 v^2$) as ρ_0 decreases with z, v^2 must increase. The factor of 2 in the exponential form occurs since, to keep $\rho_0 v^2$ constant, v only needs to e-fold over a height interval $2H$ when ρ_0 decreases by a factor of e in the height interval H. The classical observation supporting this result is shown in Fig. 3.13a. The dashed curve shows the mean wind. The actual wind fluctuates about this mean with an amplitude that increases with height. In Fig. 3.13b, a detrended version of the same data is given after subtracting the mean and multiplying by $\exp[(112 - z)/2H]$ where z and H are measured in kilometers. A schematic representation of the effect is shown in Fig. 3.13c. Many data sets show that the fluctuating component of the neutral wind velocity increases with increasing height. Furthermore, the perturbations about the average profile are often comparable to the mean wind. The altitude dependence of the Brunt-Väisälä frequency is presented in Fig. 3.14.

There is an interesting analogous behavior in waves generated by earthquakes. When a seismic wave penetrates a lower density medium such as a landfill area in the Marina District of San Francisco, to conserve energy flux in the low density region, V^2 must increase. This explains the greater damage that occurred there in the Loma Prieta quake.

Turning to the tidal case, complications arise in the analysis because the solutions are required to satisfy boundary conditions on the spherical earth. The solutions must exhibit certain altitudinal, latitudinal, and longitudinal forms referred to as Hough functions. Which propagating tidal normal modes actually are generated depends on how well the forcing function, for instance, solar or lunar forcing, matches the radial form of the mode structure. As alluded to earlier in the discussion of electrodynamics, the diurnal tide due to atmospheric heating is important at low latitudes but the response is small above 30° latitude. The semidiurnal forcing is smaller, but the altitude profiles of ozone and water vapor content fit the so-called (2,2) semidiurnal tidal mode quite well. The local heating due to these minor constituents thus couples well to the semidiurnal tide, explaining its importance at higher latitudes (Fig. 3.11).

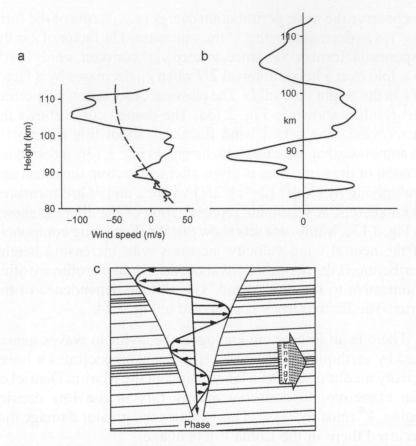

Fig. 3.13 (a) Wind components at meteor levels in a vertical plane in one representative case, derived by Liller and Whipple (1954) from the distortion of a long-enduring meteor trail. (b) Normalized wind profile at meteor heights, measured to the right and to the left from the "0" position, deduced from (a) by removal of the general shear and reduction of the residual by a factor proportional to $\rho_0^{1/2}$. (c) Pictorial representation of internal atmospheric gravity waves. Instantaneous velocity vectors are shown, together with their instantaneous and overall envelopes. Density variations are depicted by a background of parallel lines lying in surfaces of constant phase. Phase progression is essentially downward in this case, and energy propagation obliquely upward; gravity is directed vertically downward. After Hines (1974). Reproduced with permission of the American Geophysical Union.

A beautiful radar experiment was performed by Djuth et al. (1997) using plasma line observations at Arecibo. Consecutive electron density profiles are presented in Fig. 3.15. Here, we see how gravity waves can create electron density fluctuations at the level of ±1% during daytime. Notice that the phase fronts

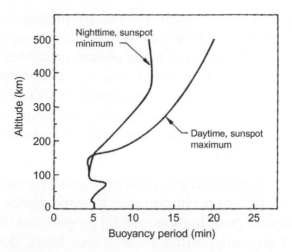

Fig. 3.14 Vertical structure of the buoyancy period computed for a standard (nonisothermal) atmosphere. Adapted from Yeh and Liu (1974). Reproduced with permission of the American Geophysical Union.

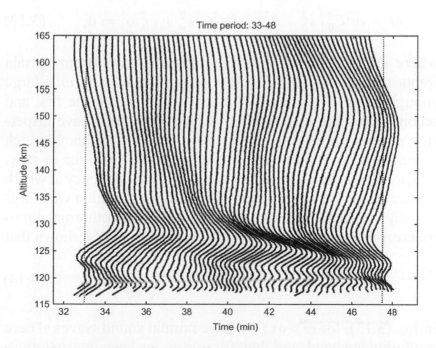

Fig. 3.15 Consecutive plasma density profiles using the plasma line technique. After Djuth et al. (1997). Reproduced with permission of the American Geophysical Union.

move downward, even though the energy propagates upward. This is a key feature of IGWs as well as tidal waves and causes the downward progression of many plasma processes described above. Notice also that these features tend to disappear above 110 km due to, as we shall see, the rapidly increasing viscosity with altitude. A longer vertical wavelength fluctuation still can be detected around 130 km, consistent with weaker viscous damping for smaller k_z.

An important implication of this exponential growth effect is that waves of little or no importance to tropospheric or stratospheric dynamics grow to monumental proportions in the E and F regions. Eventually, either these waves break down nonlinearly due to their large amplitude or the kinematic viscosity (μ/ρ_0) gets so large (as ρ_0 decreases) that viscous dissipation balances growth and eventually destroys the wave.

Equation (3.12) can also be written in the form

$$\omega^4 - \omega^2 C_0^2 \left(k_y^2 + k_z^2 \right) + \omega_b^2 C_0^2 k_y^2 + \omega_a^2 \omega^2 = 0, \qquad (3.13)$$

where $\omega_b^2 = (\gamma - 1)g^2/C_0^2$ is the square of the Brunt-Väisälä frequency and $\omega_a^2 = C_0^2/H^2$. At a given value of k_y, if ω is large enough (the high-frequency branch with $\omega > \omega_a$), the first and second terms dominate and we recover the sound wave dispersion relation found above for $g = 0$. The low-frequency branch corresponds to gravity waves that propagate only for $\omega < \omega_b$. Physically, the Brunt-Väisälä frequency is the frequency at which a parcel of air oscillates about its equilibrium position when it is initially displaced from that position. For a nonisothermal atmosphere, one in which C_0^2 varies with height, it can be shown that

$$\omega_b^2 = \frac{(\gamma - 1)g^2}{C_0^2} + (g/C_0^2) \left(\frac{dC_0^2}{dz} \right). \qquad (3.14)$$

In Eq. (3.13), for $\omega > \omega_a$, we have normal sound waves. There is a forbidden band, and then for $\omega < \omega_b$ we have internal gravity waves. IGWs travel much slower than the speed of sound, which allows them to interact with winds. There is somewhat of a similarly here between the role of the plasma frequency in electromagnetics and the Brunt-Väisälä frequency in neutral dynamics. For a plasma to have propagation, you must have $\omega > \omega_p$, the plasma frequency, whereas for IGWs, $\omega < \omega_b$ is required for propagation. Both ω_p and ω_b represent natural oscillations of the medium. Representative values for the buoyancy period $T_b = (2\pi/\omega_b)$ range from 3 to 15 min. Clearly, the 1-3-h oscillations in the ionospheric parameters discussed above fall in the gravity wave branch $\tau = (2\pi/\omega) > T_b$.

An elegant way to describe the dispersion relation for $\omega \ll \omega_a$ is

$$\omega^2 = \frac{\omega_b^2 k^2}{k^2 + m^2 + (1/4H)^2}, \qquad (3.15)$$

where $k(m)$ is the horizontal (vertical) wave number and H is the scale height. The product of the vertical phase and group velocities, $(\omega/m)(d\omega/dm)$, is negative. Thus, in the important case of upward energy propagation when $d\omega/dm$ is positive, the vertical phase velocity is downward. This downward phase propagation, as we shall see in Section 3.4, explains why plasma layers in the ionosphere often move downward, at least on average. This explains the downward progression in Fig. 3.11.

3.4 WAVE GENERATION OF ELECTRIC FIELDS

Klostermeyer (1978) first pointed out that the internal wind field, δU, in a gravity wave must also drive electrical currents, δJ. Due to the finite wavelength of the gravity waves, the associated winds are not uniform in space. The divergence of the wind-driven electrical current is therefore not zero, and an electric field, δE, must build up with a wavelength equal to that of the gravity wave. This process is illustrated in Fig. 3.16. If the E region is a perfect insulator, no field-aligned currents can flow and the $\nabla \cdot J = 0$ equation may be replaced by the more restrictive equation $\delta J = 0$, where δJ has components due to the gravity wave wind and the electric field. This requirement has been used to generate the diagram below.

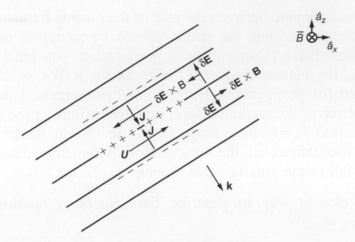

Fig. 3.16 Schematic diagram showing how the perturbation winds in a gravity wave generate electric fields. After Kelley (2009). Reproduced with permission of Elsevier.

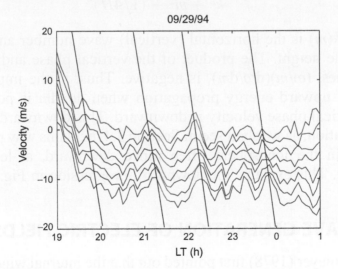

Fig. 3.17 Stack plot of the vertical drift velocities at individual heights. The lowest height is 225 km, and each subsequent line is 15 km higher and shifted by 2 m/s upward. The red portions of the plots are points inferred by interpolation. The slanted black lines are identical and have been aligned with data peaks to illustrate the downward phase velocity. (For interpretation of the references to color in this figure legend, the reader is referred to the online version of this chapter.) After Varney et al. (2009). Reproduced with permission of the American Geophysical Union.

But Prakash and Pandy (1980) showed that if **k** makes even a small angle with the plane perpendicular to **B**, the charge layers will simply short out parallel to **B**. Yet experiments seem to belie this prediction. Figure 3.17 shows that the plasma velocity due to electric fields can display downward phase velocities, as expected for neutral wind perturbations associated with an upward propagating gravity wave. To our knowledge, data such as that shown in Fig. 3.17 are the only (and definitive) examples of electric fields driven by gravity waves.

REFERENCES

Chamberlain, J.W., Hunten, D.M., 1987. Theory of Planetary Atmospheres: An Introduction to Their Physics and Chemistry, second ed. International Geophysics Series, vol. 36. Academic Press, New York.

Chapman, S., Lindzen, R.S., 1970. Atmospheric Tides: Thermal and Gravitational. D. Reidel Press, Dordrecht, Holland.

Davis, T.N., Burrows, K., Stolarik, J.D., 1967. A latitude survey of the equatorial electrojet with rocket-borne magnetometers. J. Geophys. Res. 72 (7), 1845–1861.

Djuth, F., Sulzer, M., Elder, J., Wickwar, V., 1997. High-resolution studies of atmosphere–ionosphere coupling at Arecibo Observatory, Puerto Rico. Radio Sci. 32 (6), 2321–2344.

Forbes, J., Lindzen, R.S., 1976. Atmospheric solar tides and their electrodynamic effects. Part II: The equatorial electrojet. J. Atmos. Terr. Phys. 38, 911–920.

Fritts, D.C., Isler, J.R., 1994. Mean motions and tidal and two-day structure and variability in the mesosphere and lower thermosphere over Hawaii. J. Atmos. Sci. 51 (14), 2145–2164.

Hines, C.O., 1974. The Upper Atmosphere in Motion: A Selection of Papers with Annotation. Geophysical Monographs, vol. 18. American Geophysical Union, Washington, DC.

Kelley, M.C., 1989. The Earth's Ionosphere: Plasma Physics and Electrodynamics, International Geophysics Series, vol. 43. Academic Press, San Diego, CA.

Kelley, M.C., 2009. The Earth's Ionosphere: Plasma Physics and Electrodynamics, second ed. International Geophysics Series, vol. 96. Academic Press, Burlington, MA.

Klostermeyer, J., 1978. Nonlinear investigation of the spatial resonance effect in the nighttime equatorial F-region. J. Geophys. Res. 83 (A8), 3753–3760.

Larsen, M.F., 2002. Winds and shears in the mesosphere and lower thermosphere: results from four decades of chemical release wind measurements. J. Geophys. Res. 107 (A8), 1215. http://dx.doi.org/10.1029/2001JA000218.

Larsen, M.F., Odom, C.D., 1997. Observations of altitudinal and latitudinal E region neutral wind gradients near sunset near at the magnetic equator. Geophys. Res. Lett. 24 (13), 1711–1714.

Liller, W., Whipple, F.L., 1954. High altitude winds by meteor-train photography. Rocket Exploration of the Upper Atmosphere. Pergamon Press, New York, pp. 72–130.

Matsushita, S., 1969. Dynamo currents, winds, and electric fields. Radio Sci. 4 (9), 771–780.

Maynard, N.C., 1967. Measurements of ionospheric currents off the coast of Peru. J. Geophys. Res. 72 (7), 1863–1875.

Pfaff Jr., R.F., Acuña, M.H., Marionni, P.A., Trivedi, N.B., 1997. DC polarization electric field, current density, and plasma density measurements in the daytime equatorial electrojet. Geophys. Res. Lett. 24 (13), 1667–1670.

Prakash, S., Pandy, R., 1980. On the production of large scale irregularities in the equatorial F region. In: 6th International Symposium on Equatorial Aeronomy, pp. 3–7.

Richmond, A.D., 1973a. Equatorial electrojet, I, Development of a model including winds and instabilities. J. Atmos. Terr. Phys. 35, 1083–1103.

Richmond, A.D., 1973b. Equatorial electrojet, II, Use of the model to study the equatorial ionosphere. J. Atmos. Terr. Phys. 35, 1105–1118.

Richmond, A.D., et al., 1980. An empirical model of quiet-day ionospheric electric fields at middle and low latitudes. J. Geophys. Res. 85 (A9), 4658–4664. http://dx.doi.org/10.1029/JA085iA09p04658.

Shuman, B.M., 1970. Rocket measurement of the equatorial electrojet. J. Geophys. Res. 75 (19), 3889–3901.

Varney, R.H., Kelley, M.C., Kudeki, E., 2009. Observations of electric fields associated with internal gravity waves. J. Geophys. Res. 114, A02304. http://dx.doi.org/10.1029/2008JA013733.

Yeh, K.C., Liu, C.H., 1974. Acoustic-gravity waves in the upper atmosphere. Rev. Geophys. 12 (2), 193–216. http://dx.doi.org/10.1029/RG012i002p00193.

CHAPTER 4

Electric Fields Generated by Solar Wind Interaction with the Magnetosphere

The Earth's Electric Field. http://dx.doi.org/10.1016/B978-0-12-397886-8.00004-1 87

The supersonic solar wind compresses the magnetic field of the earth on the sunward side and elongates it in the opposite direction. In the rest frame of the planet, about 1 million volts are generated across the 20 Re cross section of the magnetosphere. The electrical interaction efficiency with the earth depends on the interplanetary magnetic field (IMF) geometry. When the IMF is oppositely directed to the earth's magnetic field at the subsolar point, about 10% of this voltage enters the magnetosphere and powers all known magnetospheric processes, including the aurora, energizing most of the radiation belts, magnetic storms, million-ampere currents in the auroral zone, atmospheric winds, etc. These phenomena cause several severe space weather processes that impact humanity, including satellite disruption and communication, navigation, and power grid outages. When the magnetic fields are parallel to each other, the energy input drops but, interestingly, new phenomena arise. In this chapter, we discuss these interactions and their associated electric fields.

4.1 CONNECTION BETWEEN THE IMF AND THE EARTH'S MAGNETIC FIELD FOR B_Z SOUTH

Figure 4.1 is a cartoon showing how the two fields connect to each other for "southward" IMF. Dungey (1961) first proposed this geometry. Once connected, the illustrated field line follows the sequence as numbered, finally reconnecting far downstream in the magnetic "tail." The resulting shape looks like a comet. The voltage across the polar cap is equal to the interplanetary electric field in the earth's reference frame times the length of the connection line.

The connection zone is roughly 10% of the size of the magnetopause (the region facing the sun) in the y-direction (as shown in Fig. 4.1) and the associated voltage ranges from 30 to 100 kV. If we balance the solar wind pressure with the magnetic pressure ($B^2/2\mu_0$), the standoff distance of the magnetopause from earth is about 10 earth radii and varies only

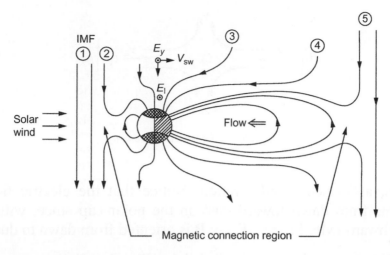

Fig. 4.1 Illustration of the connection and reconnection geometries. The upstream shock is not shown. After Kelley (2009). Reproduced with permission of Elsevier.

as 1/6th power of the solar wind pressure, so this interaction zone does not vary much in radial distance. Since the solar wind is supersonic and also super-Alfvénic, the low-frequency speed of light in a magnetized plasma (see Chapter 6), there must be a shock wave upstream of the magnetopause (not shown in Fig. 4.1) to inform the wind plasma that an object is ahead. Between the shock wave and the magnetopause, the plasma is heated and slows down. If no IMF existed, a bullet-shaped cavity would form, but very little electrical activity would result.

The length of the tail can be estimated as follows. We take the solar wind velocity, 400-1000 km/s, and multiply by the time required for a connected field line to cross the polar cap. At the earth, the electric field strength in the polar cap is the voltage generated by the connection process divided by the size of the polar cap, roughly 5000 km, yielding 10 V/km or 10 mV/m. The corresponding $\mathbf{E} \times \mathbf{B}/B^2$ velocity is then about 200 m/s. At this speed, the polar cap is traversed in 25,000 s and the length of the tail is about 2000 earth radii. When the field line reconnects, the ionosphere is no longer connected directly to the solar wind, a configuration referred to as "open." When reconnected, the field lines enter the planet at both

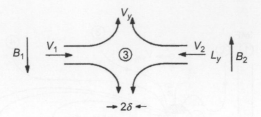

Fig. 4.2 Idealized connection geometry.

ends, a "closed" configuration. Notice that the electric field points from dawn toward dusk in the polar cap since, with a southward IMF, $\mathbf{E_{sw}} = -\mathbf{E_{sw}} \times \mathbf{B}$ is oriented from dawn to dusk.

The connection can be "explained" using the cartoon in Fig. 4.2. For simplicity, we take straight field lines on both sides to be moving toward each other at velocity $V_1 = V_2 = V_{sw}$ and with equal and opposite magnetic fields. In an infinitely conducting medium, which is a good approximation for the magnetized solar wind plasma far from the earth, the electric field is zero everywhere in the moving frame of rest and no connection would occur if this condition persisted. But in the plane where the magnetic field vanishes, the fact that the conductivity is actually finite, an electric field can exist. With the geometry shown using $\nabla \times \mathbf{B} = \mu_0 \mathbf{J}$, an electrical current must separate the two regions. For a finite conductivity, the electric field is then

$$\mathbf{E} = (\mathbf{J}/\sigma).$$

To conserve mass, the flow into the E-field zone must match the flow out of the region. Due to the geometry ($L_y \gg 2\delta$), the outflow velocity is much larger than the inflow velocity. This inflow velocity equals the $\mathbf{E} \times \delta\mathbf{B}$ velocity where $\delta\mathbf{B}$ is the small component of $\mathbf{B}$ in the x-direction and $\mathbf{E}$ is out of the paper in the dawn-dusk direction in this coordinate system. As shown in a steady state, as in Fig. 4.3a, the bend in the magnetic field is like a standing Alfvén wave moving toward the connection zone at the same velocity as the outflow velocity,

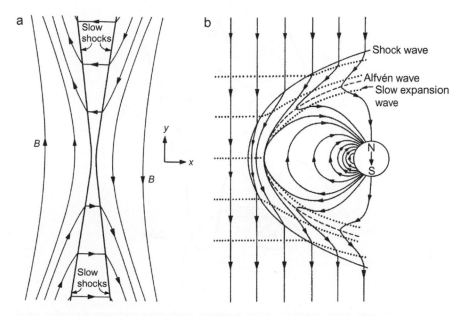

Fig. 4.3 (a) Another schematic of connection geometry. (b) Schematic diagram of how the connected field lines can maintain the same steady-state geometry by a standing Alfvén wave propagating toward the connection point at the same speed as the outflow toward the tail.

here called slow shock waves. A more realistic geometry is shown in Fig. 4.3b.

Observations show that the voltage induced across the polar cap and the currents flowing across the conducting ionosphere input a significant amount of energy to the earth's atmosphere (e.g., Joule heating, particle acceleration). But where does the energy come from to drive these processes? Figure 4.4 shows the most likely energy source. The solar wind-induced E field maps to the conducting ionosphere where a Pedersen current flows parallel to E. Since $J \cdot E > 0$, the ionosphere is a load on the system and dissipates electrical energy, but what is the source? Notice the direction of the field-aligned currents, ($J_{\parallel}$), which flow into the ionosphere at dawn and out at dusk. These currents are referred to as a component of region 1, the high-latitude currents into the ionosphere at dawn and out of the ionosphere at dusk. Region 2 currents, which close the loops of current, are discussed in Chapter 5. In the solar

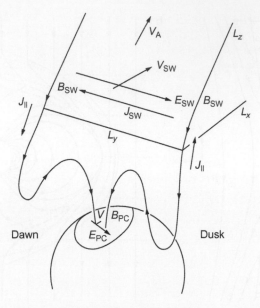

Fig. 4.4 Schematic representation of the magnetic connection between the solar wind dynamo and the ionospheric load. Note that $\mathbf{J} \times \mathbf{B}$ in the solar wind is toward the sun. After Kelley (2009). Reproduced with permission of Elsevier.

wind, these parallel currents close in the dusk-to-dawn direction such that $\mathbf{J} \cdot \mathbf{E} < 0$ and are therefore a source of electrical energy. Closure of electrical currents across a magnetic field in a collisionless plasma is accomplished by using the following equation:

$$\mathbf{J}_\perp = \frac{\mathbf{B} \times \nabla \mathbf{p}}{B^2} + \frac{\rho(\mathbf{g} \times \mathbf{B})}{B^2} - \frac{\rho\left(\frac{d\mathbf{V}}{dt} \times \mathbf{B}\right)}{B^2}.$$

Since $\nabla \mathbf{p}$ and $\mathbf{g}$ are small, the $d\mathbf{V}/dt$ term dominates. Thus, if the solar wind slows down on field lines connected to the polar cap, $\mathbf{J}$ will be dusk to dawn, as required for an energy source. Thus, mechanical energy in the flow is converted to electrical energy, which is dissipated in the ionosphere-atmosphere system. Also notice that the $\mathbf{J} \times \mathbf{B}$ force acts backward toward the sun and hence causes the slowdown in the solar wind speed. The ionospheric conductivity plays a role in this process since the current system closes in the ionospheric E region. The system was studied in detail by Hill (1984) and Siscoe et al. (2002), who derived the following formula for the polar cap potential (ϕ_{PC} or PCP):

$$\phi_{PC} = \frac{57.6 P_{sw}^{1/3} E_{sw} F(\theta)}{P_{sw}^{1/2} + 0.0125 \xi \Sigma E_{sw} F(\theta)}.$$

Here, $P_{sw}(E_{sw})$ is the pressure (electric field) in the solar wind, $F(\theta)$ and ξ are geometry factors, and Σ is the height-integrated solar cap conductivity.

Another way to think about this issue stems from the Poynting Theorem. Figure 4.5 shows how this works. Here, two

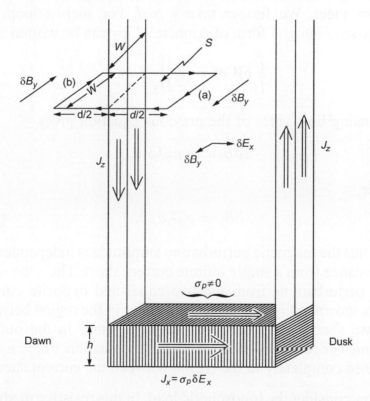

Fig. 4.5 Schematic representation of field-aligned currents and closure currents in the polar cap ionosphere and the electric and magnetic fields associated with them. After Kelley (2009). Reproduced with permission of Elsevier.

parallel current sheets of thickness dx are oriented in the y-z plane and carry equal and opposite current densities in the z-direction along the magnetic field. These current sheets represent the currents connecting the MHD generator in the solar wind to the ionosphere on the dawn and dusk sides of the polar cap, respectively, in Fig. 4.4. The current sheets are closed in the polar cap ionosphere, represented here as a resistive medium of vertical extent h having a uniform conductivity, σ_P, perpendicular to the current sheets. For this simple geometry, the height-integrated conductivity is just $\Sigma_P = \sigma_P h$. Associated with the current sheets are an extra magnetic field ($\delta\mathbf{B}$) and the impressed electric field ($\delta\mathbf{E}$). Now consider a surface S_1 bounded by a rectangular loop 1, encompassing a width w in the current sheet and extending a distance $d/2$ in each direction perpendicular to the current sheet. We further take $w \gg d$. For such a loop, the steady-state integral form of Ampère's Law can be written as

$$\oint_1 \delta\mathbf{B}\cdot d\mathbf{l} = \mu_0 \iint_{S_1} \mathbf{J}\cdot d\mathbf{a}.$$

Evaluating both sides of the previous equation gives

$$2\delta B_y w = \mu_0 J_z w d.$$

Hence,

$$\delta B_y = \mu_0 J_z d/2. \tag{4.1}$$

Note that the magnetic perturbation amplitude is independent of the distance from a single infinite current sheet. Thus, the magnetic perturbations from the two equal and opposite current sheets shown in Fig. 4.5 will add together in the region between the two sheets and exactly cancel each other in the outside regions. The result will be a magnetic perturbation, $\delta B_y = \mu_0 J_z d$, confined completely to the region between the current sheets.

Now consider the ionospheric load. In this resistive medium, the horizontal current per unit length in the y-direction from dawn to dusk is $J' = \sigma_P \delta E_x h$. In a steady state, the current per unit length entering the ionosphere in parallel current sheet 1, $J_z dx$, must be equal to the total horizontal current in the vertical extent h of the ionosphere. Thus, making use of $\delta B_y = \mu_0 J_z dx$ and the fact that $\Sigma_P = \sigma_P h$, we have

$$\delta B_y = \mu_0 \Sigma_P \delta E_x. \tag{4.2}$$

This equation may be rewritten in the equivalent forms

$$\delta E_x = \delta B_y / \mu_0 \Sigma_P, \tag{4.3}$$

$$\delta B_y / \delta E_x = \mu_0 \Sigma_P, \tag{4.4}$$

or

$$\Sigma_P = \delta B_y / \mu_0 \delta E_x. \tag{4.5}$$

Further insight comes from considering the Poynting flux. In the geometry of Fig. 4.5, $\delta \mathbf{E} \times \delta \mathbf{H}$ is downward between the current sheets and the energy input is thus $(\delta E_x \delta B_y / \mu_0)$ W/m^2. This energy must be dissipated as Joule heat in the ionosphere at the rate of $\mathbf{J} \cdot \mathbf{E} = \sigma_P \delta E_x^2$ W/m^3. Integrating over the vertical extent of the ionosphere yields a dissipation rate of $(\Sigma_P \delta E_x^2)$ W/m^3. Since the Poynting flux yields the power flow into the region per unit area, we may equate the two expressions, and once again we have the result:

$$\delta B_y / \delta E_x = \mu_0 \Sigma_P.$$

The two approaches are therefore self-consistent.

To summarize, mechanical energy is converted into electromagnetic energy in the solar wind generator. It flows down the magnetic field lines to the ionosphere as Poynting flux, where it is converted into heat by Joule dissipation. For the typical parameters of $\delta E_x = 50$ mV/m and $\delta B_y = 500$ nT, we can estimate the Poynting flux to be $\delta E_x \delta B_y / \mu_0 = 0.02$ W/m^2 = 20 ergs/cm^2 s. This is a substantial amount of energy, roughly 10^{11} W over the whole region. It is important to notice also that an energy flux of 20 ergs/cm^2 s is very large compared to typically observed energy fluxes in auroral particle precipitation except for extremely intense localized auroral arcs. In fact, the Joule heat input is the primary reason that the thermosphere has a local temperature maximum in the high-latitude region, which competes with the solar photon-driven temperature maximum at the subsolar point.

The stretched magnetic field has much more energy than a simple dipole, and it is this energy that powers all known auroral processes. In a steady state, as much energy is input at the front of the earth by the connection process as is dissipated in tail reconnection. This system is unstable, however, and occasionally releases huge amounts of energy in a 1-h time frame called a magnetic substorm. Each time this happens, the magnetic field becomes more like a dipole, which is the ground state of the system. When B_z stays south for a long time, a series of substorms occur, resulting in a magnetic storm. A substorm can change the magnetic field locally in the auroral zone by as much as 4%. A magnetic storm can change the field globally by 1%.

The time variations associated with these magnetic fields are a serious problem for electrical transformers due to geomagnetically induced currents (Kappenman, 2003, 2005). Quebec was out of power for 10 h at a cost of $10 billion during a 1989 storm. Even South Africa has had such disruptions. The worldwide output of transformers could not make up for a major storm for years, so this is a very serious problem. Time-varying auroral currents can generate a dB/dt of 60 nT/min. Even the equatorial electrojet can generate 5 nT/min during events when the solar wind or magnetospheric electric field penetrates to the equator (Ilma et al., 2012).

Satellites crossing the polar cap can be used to measure the potential as shown in Fig. 4.6. Flow across the polar cap merges with the auroral zone flow to form a two-cell convection pattern as shown in Fig. 4.7 for various IMF B_y configurations with B_z southward.

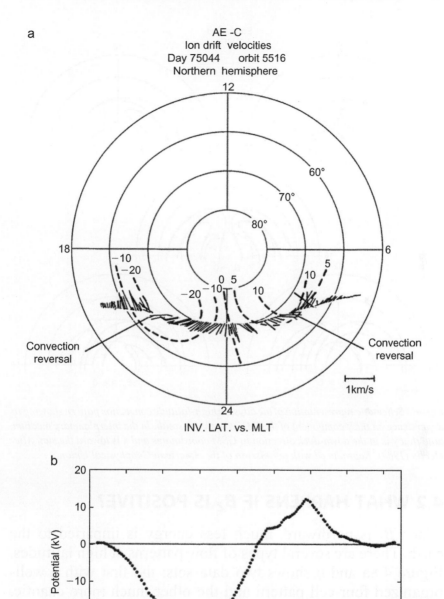

a

AE -C
Ion drift velocities
Day 75044   orbit 5516
Northern hemisphere

Convection
reversal

Convection
reversal

1km/s

INV. LAT. vs. MLT

b

Fig. 4.6 (a) A satellite flight across the high-latitude convection pattern provides drift velocity profiles, which are shown along with the inferred convection pattern. (b) The potential distribution resulting from this convective flow pattern shows maxima and minima at the polar cap boundaries and an average total potential difference of about 60 kV across the polar cap. After Heelis and Hanson (1980). Reproduced with permission of the American Geophysical Union.

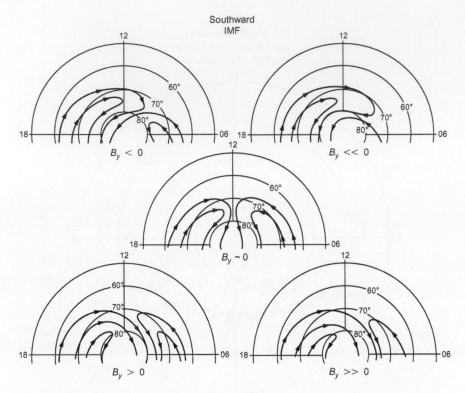

Fig. 4.7 Schematic representation of the dayside high-latitude convection pattern showing its dependence on the y component of the IMF when B_z is south. In the interplanetary medium, note that y is in the dawn-dusk direction in GSM coordinates and x is toward the sun. After Heelis (1984). Reproduced with permission of the American Geophysical Union.

4.2 WHAT HAPPENS IF B_Z IS POSITIVE?

When B_z is northward, much less energy is imparted to the earth. There are several types of flow patterns at high latitudes. Figure 4.8a and b shows two data sets: the first with a well-organized four-cell pattern and the other much more chaotic. We consider the first case first. At low latitudes corresponding to a shrunken auroral oval, conventional wisdom is that viscous interaction is the source, that is, the solar wind flow around the earth grabs plasma and viscously pulls it toward and away from the sun, as illustrated in Fig. 4.9. Note that if the plasma is moving across a magnetic field, $\mathbf{E} = -\mathbf{V} \times \mathbf{B}$ and a viscous-induced flow yields an electric field.

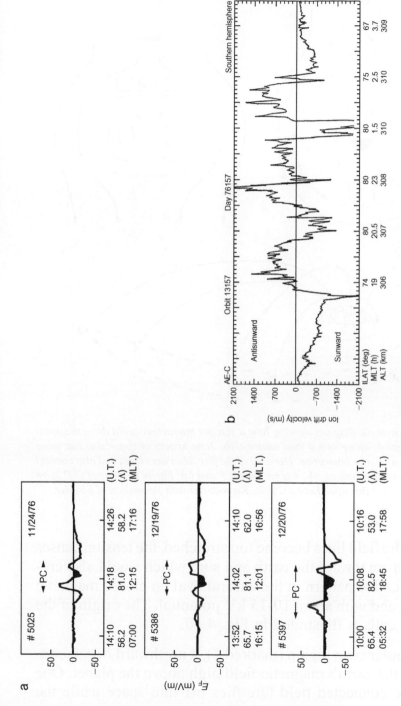

Fig. 4.8 (a) Three examples of the high-latitude, dawn-to-dusk component of the electric field in the ionosphere for a northward IMF. The shaded regions indicate sunward plasma flow. (b) High-latitude plasma drift velocity in the ionosphere shown for a northward IMF. The flow is extremely structured and does not indicate a simple two-cell convection pattern, as is often the case during southward IMF. Panel (a): After Burke et al. (1979). Reproduced with permission of the American Geophysical Union. Panel (b): After Heelis and Hanson (1980). Reproduced with permission of the American Geophysical Union.

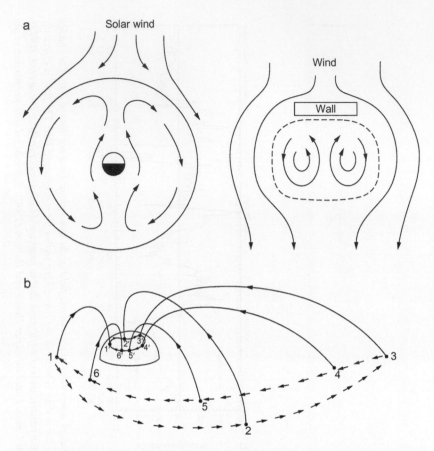

Fig. 4.9 (a) Schematic diagram showing how a viscous interaction could drive magneto-spheric convection, along with a fluid analogy. (b) Time history of convecting flux tubes resulting from a viscous interaction. Here, all of the flux tubes are closed and also connect into the Southern Hemisphere, which is not shown. Panel (a): Figure courtesy of D. P. Stern and S. Lantz. Panel (b): After Kelley (2009). Reproduced with permission of Elsevier.

When the field lines become too stretched, the tension causes them to return toward the earth and sun, which would also create a two-celled pattern in the auroral oval on closed magnetic field lines and with about 10-15 kV potential. The origin of the other two cells is illustrated in Fig. 4.10.

In the polar cusp region (marked 1), a northward B_z can connect with the earth's magnetic field high above the planet. One end of the connected field line flies off into space while the

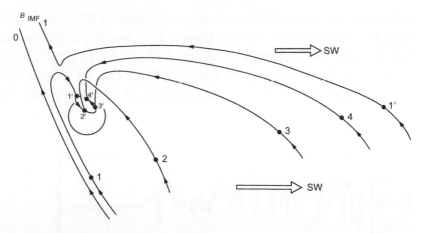

Fig. 4.10 Time history of convecting flux tubes (connecting to a northward IMF) that pro-duce a dominant high-latitude convection cell. This must be combined with the pattern for viscous interaction shown in Fig. 4.9 to produce the observed four-cell pattern. After Kelley (2009). Reproduced with permission of Elsevier.

other is dragged across the pole. Again, once stretched, it rebounds toward the sun in the center of the polar cap and we have a four-cell pattern. This phenomenon is the origin of the residual PCP in Fig. 4.1 for B_z north. The Hill-Siscoe model only treats dayside B_z south physics.

The situation in Fig. 4.8b is more complicated and interesting. The variations are suggestive of two-dimensional turbulence (Kraichnan, 1967), as discussed by Kelley and Kim (2012).

4.3 PENETRATION OF SOLAR WIND ELECTRIC FIELDS THROUGHOUT THE INNER MAGNETOSPHERE

Remarkably, the solar wind electric field penetrates deep into the magnetosphere, even to the equatorial ionosphere and down to stratospheric heights (Kelley, 2009; Kelley et al., 1979). For B_z south, the efficiency is 10%, that is, for a 30 mV/m interplanetary electric field that is primarily in the dawn-to-dusk direction (called IEFy), 3 mV/m reaches deep into the system. The system acts like a high-pass filter such that variations with periods shorter than a few hours penetrate the entire system. An example is given in Fig. 4.11. Here, IEFy is plotted in the top panel as measured on the ACE satellite at the L1 libration point. In the middle, it is divided by 10

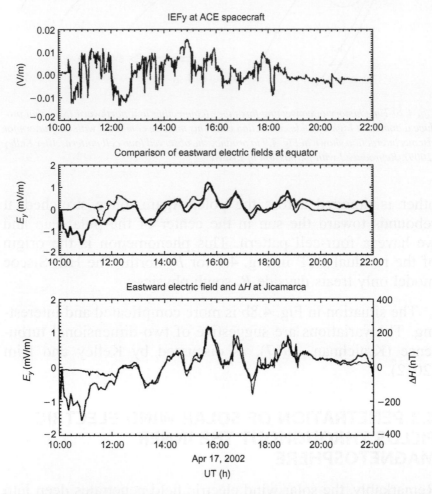

Fig. 4.11 Direct comparison of the IEF and the equatorial electric field (using ISR and the Anderson technique) after adjusting for the time delay from the ACE satellite to the magnetopause. In the middle and bottom panels, the dark line is the eastward component over Jicamarca corresponding to the dawn-to-dusk direction. The light line in the middle panel is the dawn-to-dusk component of the IEF. In the lower panel, the light line is the predicted eastward field from H values. After Kelley et al. (2003). Reproduced with permission of the American Geophysical Union.

and delayed by the transit time (Kelley and Dao, 2009) and plotted on top of the eastward component of the electric field measured at the equator. The correlation is over 90%. Below, the IEFy is compared to the magnetic field measured on the ground in Peru. Penetration occurs for B_z north as well but at the 3% level (Kelley et al., 2004). Such events can drastically affect the potential between dawn and dusk, typically 12 kV, changing its sign and tripling it throughout the low-latitude atmosphere.

The connection of this voltage pattern to atmospheric electricity can be seen from Fig. 4.12. Using the fact that the line integral of the electric field must vanish around any closed loop, we choose a closed contour with one path through the equatorial ionosphere and a parallel path inside the conducting earth. The contour is completed vertically at dusk and dawn. Typically, the earth-ionosphere potential difference due to the fair weather electric field is the order of 250 kV, but since a finite potential difference exists between dusk and dawn in the ionosphere but no similar potential in the path through the earth, the potential between the earth and the ionosphere cannot be the same at dusk and dawn. For the quiet time electric field, there is a few percent difference in the earth-ionosphere potential, but during penetrating electric field events, this difference can increase to over 10%. In turn, this implies a 5% difference in the air-earth current density (J_z) with the larger current at dawn.

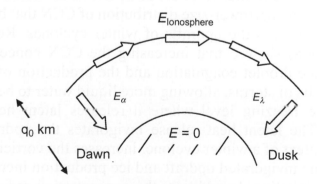

Fig. 4.12 Line integral discussed in the text.

Numerous papers, starting with Roberts and Olson (1973) and Wilcox et al. (1973) and persisting in later updates (see the review by Tinsley, 2008), show highly significant correlations on the day-to-day timescale of small vorticity changes in winter storms with atmospheric inputs modulated by the solar wind. The inputs include Forbush decreases of the cosmic ray flux, the precipitation of relativistic electrons into the stratosphere, and the precipitation of solar energetic particles into the polar caps. In addition, Burns et al. (2007, 2008) showed that not only do changes in the IMF B_y, a proxy for the IEF, produce surface pressure changes inside the polar vortices correlated with J_z changes, but similar and larger responses are also produced there in response to J_z changes from day-to-day changes in global thunderstorm activity. All of the atmospheric dynamic responses can be understood as responses to changes in J_z that affect cloud microphysics through the deposition of electric charge on droplets and aerosol particles that affect the scavenging rates by droplets in clouds of both cloud condensation nuclei (CCN) and ice-forming nuclei. This phenomenon can be termed "charge modulation of aerosol scavenging" (CMAS) and is due to the long-range repulsive Coulomb force reducing Brownian scavenging, thereby increasing the concentration of small CCN, and the short-range image attractive force increasing scavenging and decreasing the concentration of large and giant CCN (Tinsley, 2010).

Over the course of a day or so in cloudy, winter oceanic environments, the cumulative effect of CMAS is an increased concentration and narrower-size distribution of CCN that becomes incorporated into the updrafts of winter cyclones. Rosenfeld et al. (2008) showed that increasing the CCN concentration reduces the droplet coagulation and the production of rain in the updrafts of storms, allowing more liquid water to be carried above the freezing level where it releases latent heat as it freezes. The latent heat release invigorates the updraft and, in the context of a winter cyclone, increases the vorticity. Ultimately, the invigorated updraft and ice production increase the total precipitation. In addition, the concentration reduction of large and giant CCN by CMAS reduces the concentration of large droplets that initiate the coagulation processes and contribute to the increased transport of water above the freezing level. This mechanism accounts for the observed correlations of winter storm vorticity with J_z and solar cycle effects on blocking (Tinsley, 2012).

In the context of tropical deep convection, which is relevant to equatorial penetrating solar wind electric field effects, the model of Chen and Yin (2011) shows that the same process, responding to increases in CCN concentration, increases the transport of ice and, thus, water vapor into the lower stratosphere.

While we are unaware of observations showing tropical deep convection responses to magnetic storms, evidence for such responses in cloud parameters and water vapor concentration in the lower stratosphere would be worth looking for.

REFERENCES

Burke, W.J., Kelley, M.C., Sagalyn, R.C., Smiddy, M., Lai, S.T., 1979. Polar cap electric field structures with a northward interplanetary magnetic field. Geophys. Res. Lett. 6 (1), 21–24. http://dx.doi.org/10.1029/GL006i001p00021.

Burns, G.B., Tinsley, B.A., Frank-Kamenetsky, A.V., Bering, E.A., 2007. Interplanetary magnetic field and atmospheric electric circuit influences on ground level pressure at Vostok. J. Geophys. Res. 112, D04103.

Burns, G.B., Tinsley, B.A., French, W.J.R., Troshichev, O.A., Frank-Kamenesky, A.V., 2008. Atmospheric circuit influences on ground level pressure in the Antarctic and Arctic. J. Geophys. Res. 113, D15112.

Chen, B., Yin, Y., 2011. Modeling the impact of aerosols on tropical overshooting thunderstorms and stratospheric water vapor. J. Geophys. Res. 116, D19203. http://dx.doi.org/10.1029/2011JD015591.

Dungey, J.W., 1961. Interplanetary magnetic field and the auroral zones. Phys. Rev. Lett. 6 (2), 47–48.

Heelis, R.A., 1984. The effects of interplanetary magnetic field orientation on dayside high-latitude ionospheric convection. J. Geophys. Res. 89 (A5), 2873–2880. http://dx.doi.org/10.1029/JA089iA05p02873.

Heelis, R.A., Hanson, W.B., 1980. High latitude ion convection in the nighttime F-region. J. Geophys. Res. 85 (A5), 1995–2002. http://dx.doi.org/10.1029/JA085iA05p01995.

Hill, T.W., 1984. Magnetic coupling between solar wind and magnetosphere: regulated by ionospheric conductance? (abstract). Eos Trans. AGU 65 (45), 1047.

Ilma, R.R., Kelley, M.C., Gonzales, C.A., 2012. On a correlation between the ionospheric electric field and the time derivative of the magnetic field. Int. J. Geophys. 2012, 648402. http://dx.doi.org/10.1155/2012/648402.

Kappenman, J.G., 2003. Storm sudden commencement events and the associated geomagnetically induced current risks to ground-based systems at low-latitude and midlatitude locations. Space Weather 1 (3), 1016. http://dx.doi.org/10.1029/2003SW000009.

Kappenman, J.G., 2005. An overview of the impulsive geomagnetic field disturbances and power grid impacts associated with the violent Sun–Earth connection events of 29–31 October 2003 and a comparative evaluation with other contemporary storms. Space Weather 3, S08C01. http://dx.doi.org/10.1029/2004SW000128.

Kelley, M.C., 2009. The Earth's Ionosphere: Plasma Physics and Electrodynamics, second ed. International Geophysics Series, vol. 96. Academic Press, Burlington, MA.

Kelley, M.C., Dao, E., 2009. On the local time dependence of the penetration of solar wind-induced electric fields to the magnetic equator. Ann. Geophys. 27, 3027–3030. http://dx.doi.org/10.5194/angeo-27-3027-2009.

Kelley, M.C., Kim, H.-J., 2012. A suggestion that two-dimensional turbulence contributes to polar cap convection for B_z north. Geophys. Res. Lett. 39 (A1), L07102. http://dx.doi.org/10.1029/2012GL0513472012.

Kelley, M.C., Fejer, B.G., Gonzales, C.A., 1979. An explanation for anomalous ionospheric electric fields associated with a northward turning of the interplanetary magnetic field. Geophys. Res. Lett. 6 (4), 301–304. http://dx.doi.org/10.1029/GL006i004p00301.

Kelley, M.C., Kruschwitz, C.A., Gardner, C.S., Drummond, J.D., Kane, T.J., 2003. Mesospheric turbulence measurements from persistent Leonid meteor train observations. J. Geophys. Res. 108 (D8), 8454. http://dx.doi.org/10.1029/2002JD002392.

Kelley, M.C., Wong, V.K., Hajj, G.A., Mannucci, A.J., 2004. On measuring the off-equatorial conductivity before and during convective ionospheric storms. Geophys. Res. Lett. 31, L17805. http://dx.doi.org/10.1029/2004GL020423.

Kraichnan, R.H., 1967. Inertial ranges in two-dimensional turbulence. Phys. Fluids 10, 1417–1423. http://dx.doi.org/10.1063/1.1762301.

Roberts, W.O., Olson, R.H., 1973. Geomagnetic storms and wintertime 300 mb trough development in the North Pacific–North America area. J. Atmos. Sci. 30, 135–140.

Rosenfeld, D., Lohmann, U., Raga, G.B., O'Dowd, C.D., Kulmala, M., Fuzzi, S., Reissell, A., Andreae, M.O., 2008. Flood or drought: how do aerosols affect precipitation? Science 321, 1309–1313. http://dx.doi.org/10.1126/science.1160606.

Siscoe, G.L., Erickson, G.M., Sonnerup, B.U.Ö., Maynard, N.C., Schoendorf, J.A., Siebert, K.D., Weimer, D.R., White, W.W., Wilson, G.R., 2002. Hill model of transpolar potential saturation: comparisons with MHD simulations. J. Geophys. Res. 107 (A6), 1075. http://dx.doi.org/10.1029/2001JA000109.

Tinsley, B.A., 2008. The global atmospheric electric circuit and its effects on cloud microphysics. Rep. Prog. Phys. 71 (6), 066801.

Tinsley, B.A., 2010. Electric charge modulation of aerosol scavenging in clouds: rate coefficients with Monte-Carlo simulations of diffusion. J. Geophys. Res. 115, D23211.

Tinsley, B.A., 2012. A working hypothesis for connections between electrically-induced changes in cloud microphysics and storm vorticity, with possible effects on circulation. Adv. Space Res. 50 (6), 791–805. http://dx.doi.org/10.1016/j.asr.2012,04,008.

Wilcox, J.M., Scherrer, P.H., Svalgaard, L., Roberts, W.O., Olson, R.H., 1973. Solar magnetic structure: relation to circulation of the earth's atmosphere. Science 180 (4082), 185–186.

Jacob, D.J., Crutzen, P.J., Sanhueza, E.M., Matson, P.A., Schimel, D.S., Nicholls, R.J., Watson, R.T., White, D.R., Wilson, O.R., 2002. Hill model of human-plate potential sediment comparisons with MLD simulations. J. Geophys. Res. 107 (A6), 1112. http://dx.doi.org/10.1029/2001JA001107.

De Jey, H.A., 2003. The global atmospheric electric circuit and its effect on cloud microphysics. Rep. Prog. Phys. 71 (6), 066801.

Tinsley, B.A., 2014. Electric charge modulation of aerosol scavenging in clouds: rate coefficients with Monte-Carlo simulations of diffusion. J. Geophys. Res. 115 (D23211).

Tinsley, B.A., 2012. A working hypothesis for connections between electrically induced changes in cloud microphysics and storm vorticity, with possible effects on circulation. Adv. Space Res. 50 (6), 791–805. http://dx.doi.org/10.1016/j.asr.2012.04.008.

Wilcox, J.M., Scherrer, P.H., Svalgaard, L., Roberts, W.O., Olson, R.H., 1973. Solar magnetic structure relation to circulation of the earth's atmosphere. Science 80 (4093), 185–186.

CHAPTER 5

The Earth's Magnetosphere: A Secondary Collisionless Magnetohydrodynamic Generator

The Earth's Electric Field. http://dx.doi.org/10.1016/B978-0-12-397886-8.00005-3   109

That the earth and its atmosphere are rotating is communicated along magnetic field lines by electric fields. These electric fields put the plasma in motion so, to first order, the entire inner magnetosphere also rotates with the earth. In the sun's frame of reference, this is the largest electric field source at latitudes less than about 60°, whereas in the rotating frame, this field vanishes and the processes in Chapter 3 dominate. At higher latitudes corresponding to magnetic field-line crossings at higher altitudes in the equatorial plane (but still below the polar cap), the electric field is determined by the flow field inside the magnetosphere. The flow slows down as it approaches the earth and the mechanical energy is converted to electric energy. This is accompanied by the so-called region 2 field-aligned currents and a variety of electric field generation effects. As particles approach the earth, they become energized and form the outer Van Allen radiation belts. Large $\partial B/\partial t$ electric fields can occasionally be generated at the earth's surface where they endanger electric power systems. Large polarization fields occur in many situations and strong parallel electric fields generate the primary auroral particles. As these energetic particles penetrate deeper into the magnetosphere, they cross electric potential lines, reach 300,000,000 K, and fill the outer Van Allen radiation belts. The combination of varying flow fields and plasma gradients creates many localized electric field structures.

5.1 EARTH'S ROTATION AS AN ELECTRIC FIELD GENERATOR

The earth's atmosphere, of course, rotates with the solid earth. If it did not, the wind speed at the equator would be 434 m/s, which is the velocity with which the surface of the earth moves at the equator. As long as the neutral density is well in excess of the plasma density, the ions must also corotate. This high density condition certainly holds at the height of one of the primary collisional magnetohydromagnetic generators: the E region. Above this height, collisions with neutrals become less and less, and knowledge from the plasma that the earth is even rotating at all is not obvious. However, as argued next, this information is not lost to the ionized material. If, then, we grant that at least ions are tied to neutrals in the E region, then the electrons must also follow along, and plasma in the E region must corotate with the planet. In this frame, designated by subscript "R," E_R and V_R both vanish. Now suppose that we go into a coordinate system fixed to the sun where the electric field is $-V_R$. The electron and ion velocities are both zero. The magnetic field is unchanged in such a nonrelativistic transformation, but $E_S = E_R - V_R \times B$, since in transforming to this reference frame, the coordinates move at velocity $-V_R$. However, in a nonrotating frame, one fixed with respect to the sun, there must be an electric field such that

$$E_S = -V_R \times B. \tag{5.1}$$

Thus, we have E, inward in the equatorial plane in the sun-fixed frame, and the equipotential lines in the equatorial plane are circles, as illustrated in the central portion of Fig. 5.1. To check

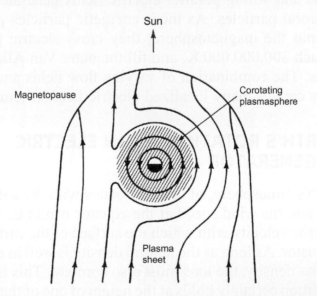

Fig. 5.1 Contours of plasma flow velocity and equipotentials in the magnetic equatorial plane. The concentric circles indicate corotating plasma deep in the magnetosphere. The diagram is fixed with respect to the sun. After Kelley (2009). Figure courtesy of R. Heelis and reproduced with permission of Elsevier.

for validity, note that the magnetic field in the equatorial plane is northward and, with $\mathbf{E_S}$ inward, $\mathbf{E_S} \times \mathbf{B}/B^2$ is eastward, so the plasma indeed rotates with the planet. This frame is important, since the solar wind and magnetospheric generators also operate in the sun-fixed frame and, in that frame, the effects add together, as shown in the figure and as discussed below. The field lines threading the plane shown are all closed and hence touch the earth in both hemispheres.

The corotating region is called the plasmasphere since it is a region of dense, cold plasma. Every day, the plasma generated in the dayside by sunlight flows out along the magnetic field lines to fill the region. At night, the flow reverses but its lifetime is such that, to first order, the whole region remains filled with cold plasma.

To gauge where the two effects—solar wind and earth rotation—become comparable, Mozer (1973) compared the corotation electric field in the sun-fixed frame with two levels of magnetospheric electric fields: 0.4 and 1 mV/m, as shown in Fig. 5.2 (as discussed in Section 5.2). The corotation field in the ionosphere is meridional in the sun-fixed frame and has the form:

$$E_{\mathrm{MI}}(\theta) = 14\cos\theta\left(1 + 3\sin^2\theta\right)^{1/2} \mathrm{mV/m}. \tag{5.2}$$

This electric field can be mapped from the ionosphere to the equatorial plane using formulas determined from the dipole magnetic field spreading, assuming $E_{\parallel} = 0$ (Mozer, 1970). This is equivalent to the notion that magnetic field lines are perfect conductors and that the voltage difference is constant between any two field lines. Thus, as the field lines separate from the ionosphere to the equatorial plane, the voltage stays the same but the electric field becomes smaller. These factors are different for meridional and zonal electric field components, that is,

$$\frac{E_{\mathrm{MI}}}{E_{\mathrm{MM}}} = 2L\left(L - \frac{3}{4}\right)^{1/2}, \tag{5.3}$$

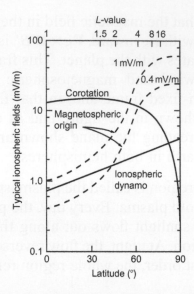

Fig. 5.2 Typical ionospheric level electric fields observed in a nonrotating frame of reference that arise from corotation, the interaction of tidal neutral winds with the ionosphere, and the interaction of the solar wind with the terrestrial magnetic field. After Mozer (1973). Reproduced with permission of the American Geophysical Union.

$$\frac{E_{ZI}}{E_{ZM}} = L^{3/2}. \qquad (5.4)$$

In Fig. 5.2, we use values at ionospheric heights. The parameter L is the distance in earth radii where a given field line crosses the equator. The corotation curve crosses the magnetospheric curves between $L = 3$ Re and $L = 5$ Re. Thus, magnetic flux tubes inside this range of L shells corotate and can fill with plasma from the dayside of the planet. The result is a dense cold plasma region called the plasmasphere inside about 20,000 km altitude. On average, this yields a toroidal-shaped plasma region with a typical density of 100 cm^{-3} at the equatorial plane and a typical temperature of 1 eV. At higher L shells, the flux tubes have huge volumes or are connected with the solar wind and hence have even larger volumes. In such cases, the plasma created by either the sun or the aurora just expands into near vacuum. This explains the low plasma densities outside the plasmasphere. Curiously, the plasma ion above about 600 km altitude inside the plasmasphere is H^{+} and outside is O^{+}. To explain the former case, we note that the reaction $H + O^+ \Leftrightarrow H^+ + O$ is very fast and, since the earth has a geocorona full of hydrogen, O^+ is replaced rapidly by H^+. In the latter case, the ionospheric O^+ ions are expelled by a "polar wind" or accelerated by plasma waves and become dominant species below several thousand kilometers.

Inside the plasmasphere, to first order, everything rotates with the earth. But it is in exactly this coordinate system that the hydromagnetic generator of Chapter 3 operates. Since atmospheric winds are usually less than 434 m/s, the corresponding electric fields are less than the corotation field in Eq. (5.2). In the nonrotating frame, these fields are a minor perturbation on the corotation field.

Things become interesting at the plasmaspheric boundary. In the nonrotating frame, we know that the magnetospheric flow magnitude equals the corotation speed at that location. Since the magnetospheric flow is sunward (see below), corotation in the dusk sector opposes this flow and stagnation occurs. The sum of the two potentials is shown schematically in Fig. 5.3. Since this stagnation occurs in virtually permanent darkness, ion chemistry (Kelley, 2009) results in a steady decrease in plasma content

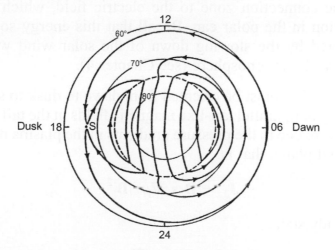

Fig. 5.3 Sum of the corotation and magnetospheric electric field in a nonrotating frame. The flow speed goes through zero at around 21:00 at 60° latitude and the flux tube is never sunlit. After Kelley (2009). Figure courtesy of R. Heelis and reproduced with permission of Elsevier.

due to recombination. The result is a deep depletion in plasma density and the so-called mid-latitude ionospheric trough develops. We return to this region in Section 5.5.

Jupiter rotates once every 10 h and the magnetic field is much larger than the earth's. Thus, the corotation-dominated region extends out to 10 R_J. Here, the limiting factor is not the magnetospheric electric field but where the huge corotation velocity equals the Alfvén speed. This speed is, in a real sense, the speed of light for low-frequency waves. This region of the Jovian magnetosphere is thus quite interesting but for a different reason than that of the earth.

5.2 MAGNETOHYDROMAGNETIC ELECTRIC FIELD GENERATION INSIDE THE MAGNETOSPHERE

In a steady state, magnetic connection at the front of the magnetosphere must be balanced on average by reconnection in the tail. By the right-hand rule, this magnetopause current must also be oriented from dawn to dusk. Thus, the electric field is parallel to the magnetopause current and $\mathbf{J} \cdot \mathbf{E} > 0$. So, this is a region of magnetic energy dissipation that results in outflow from the connection zone to the electric field, which drives convection in the polar cap. Recall that this energy source is augmented by the slowing down of the solar wind when in contact with the ionosphere (see Chapter 4).

In the tail, $\mathbf{J}$ must be oriented from dawn to dusk to support the nearly antiparallel tail-like magnetic fields in the tail. These currents are due to the pressure gradients in the plasma near the equatorial plane, that is:

$$\mathbf{J}_\perp = \mathbf{B} \times \nabla p / |\mathbf{B}|^2.$$

In a steady state,

$$\mathbf{J} = \mathbf{B} = -\nabla p,$$

or equivalently, using one of Maxwell's Equations, we have,

$$\frac{1}{\mu_0} (\nabla \times \mathbf{B}) \times \mathbf{B} = \nabla p. \tag{5.5}$$

The left-hand side of this equation is purely geometrical and can be described in terms of magnetic pressure and tension. This is the rubber band model of magnetic field lines (described in plasma physics texts), which snap back toward the earth after they reconnect on the nightside. In a steady state, magnetic field geometry, these flux tubes must be replaced by those connecting to the solar wind on the dayside. In both connection and reconnection, magnetic energy is converted to flow energy as the field lines converge on each other and disappear. The circulation that occurs is called convection, even though it is not driven by heating.

To proceed further, we need to review Poynting's Theorem (see Chapter 1), which can be written:

$$\frac{\partial}{\partial t} \iiint \left(\frac{\varepsilon E^2}{2} + \frac{\mu H^2}{2} \right) dV = -\oiint (\mathbf{E} \times \mathbf{H}) \cdot d\mathbf{s}$$

$$+ \iiint (\mathbf{J} \cdot \mathbf{E} + \mathbf{V} \cdot (\mathbf{J} \times \mathbf{B}) dV).$$

First, we compare the ratio, R, of electric and magnetic field energy densities in the magnetosphere. Using $\varepsilon = (c^2/V_A)^2$,

$$R = \frac{\mu_0 H^2}{\varepsilon E^2} = \frac{B^2}{\varepsilon_0 \mu_0 E^2} = \frac{B^2 V_A^2}{E^2}.$$

For $B = 100$ nT, $V_A = 10^6$ m/s and $E = 10^{-3}$ V/m, $R = 10^4$ and the magnetic field is dominant in terms of energy density. Below, we consider several different applications of Poynting's Theorem.

5.2.1 Steady B_z South

This has every indication of being a steady-state situation for the magnetosphere, as illustrated in Fig. 5.4, but it is actually one of great dynamics and energy transfer from the solar wind to the earth's atmosphere. Reconnection geometry for the tail is

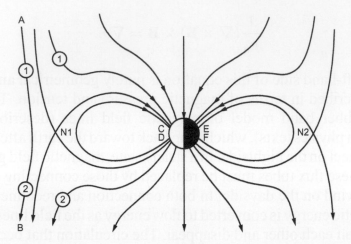

Fig. 5.4 Schematic diagram showing dayside merging (N1) and nightside reconnection (N2) in the noon-midnight meridian. After Kelley (2009). Figure courtesy of D.P. Stern and reproduced with permission of Elsevier.

also illustrated in Fig. 5.4 where the geometry has been distorted for simplicity. At the nighttime X-line (N2), analogous to connection on the dayside, there is an electric field in the region of small magnetic field. The argument for this is identical to that given in Chapter 4 for the dayside. Its direction must be from dawn to dusk so that above and below the equatorial plane, plasma moves toward that plane. Since there is a small component of the magnetic field normal to the equatorial plane, this same electric field causes convection toward the sun earthward of the X-line and tailward convection outside the X-line. The field lines reconnected to the solar wind just blow away. Here, the $(\mathbf{E} \times \mathbf{H}) \cdot \mathbf{d}^5$ term is clearly inward on closed field lines and energy is being fed into the magnetosphere. (Recall that $\mathbf{d}^5$ is outward.). Since there is no change in the magnetic field topology, $\partial B/\partial t = 0$, in this steady state, the energy must be going into ionospheric Joule heating, auroral particle acceleration, and work on the ring current plasma. Notice that $\mathbf{J} \cdot \mathbf{E} > 0$ in the reconnection region, so magnetic energy is being disrupted and converted to mechanical energy.

5.2.2 Cranking Up B_z South

Suppose now that we slowly but steadily increase connection and reconnection. From Chapter 4, we know that region 1 currents will increase (see below) but so will the region 2 return currents. Likewise, $\mathbf{J} \cdot \mathbf{E}$ will increase in the ionosphere and the magnetic tail will expand. At some point, in a way not yet totally understood, a global instability called a magnetic substorm is generated. This is initiated deep in the magnetosphere where the most equatorward auroral arcs brighten and then, over an hour or so, they surge rapidly poleward. Large zonal currents flow with examples of magnetic fields over 4% of the earth's field. An example of the aurora during such an event is presented in Fig. 5.5.

When this release of magnetic energy occurs, the magnetic field geometry tends to collapse to its lowest energy state, a dipole field. This is termed "depolarization." But if B_z stays southward for a long time, the process of energy storage and release can occur many times and the event becomes a magnetic storm. The ring current is driven by huge pressure gradients due to the buildup of energy in the particles originally energized in the aurora (see below). As particles are brought into the inner magnetosphere, the Poynting flux term $\mathbf{V} \cdot \mathbf{J} \times \mathbf{B}$ is the work done on the plasma, increasing their energy from 5 to 50 keV. Also, consider that gradient and curvature drifts move particles across potential lines. Since many tens of keV are available, the temperature reaches 300,000,000 K!

These storms are worldwide events with magnetic field decreases of as much as 2000 nT, such as during the Carrington storm of 1859, 4% of the earth's field everywhere (aurora was observed in Bombay!). In a modest storm in 1989, Quebec lost power for 10 h at a cost of $10 billion. Sweden, Finland, and even South Africa were affected during the same period. When, not if, a significantly destructive storm occurs again, major technological effects will occur, including loss

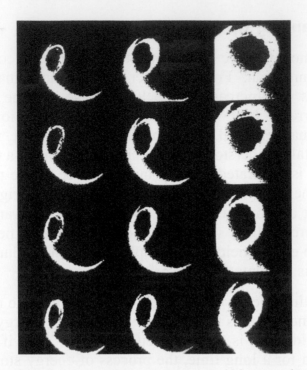

Fig. 5.5 A magnetic substorm viewed from several thousand kilometers altitude. The poleward surge is evident in the top right-hand panel. Here, 12 consecutive images at ultraviolet wavelengths 123-160 nm record the development of an auroral substorm. The sequence begins at 05:29 UT on April 2, 1982 (bottom left image) as the NASA/GSFC spacecraft Dynamics Explorer I first views the auroral oval from the late evening side of the dark hemisphere at low northern latitudes near apogee (3.65 earth radii altitude) and then from progressively greater latitudes as the spacecraft proceeds inbound over the auroral oval toward perigee. In each panel, time increases upward and from left to right. The poleward bulge at onset of the auroral substorm is observed beginning at 06:05 UT (fourth frame at upper left). In successive 12-min images, the substorm is observed to expand rapidly in latitude and longitude. After Kelley (2002). Figure courtesy of L.A. Frank, J.D. Craven, and R.L. Rairden and reproduced with permission of Elsevier.

of electrical power due to the effect of geomagnetically induced currents on transformers, which are very hard to replace (Kappenman, 2005). Ilma et al. (2012) show that, even in the tropics, more than 10 nT/min changes in the magnetic field have occurred, which we estimate could be 5-10 times greater in a Carrington storm. This is a real problem, which is not even on the radar of the world's leaders. Figure 5.6 shows the relationship between auroral zone and inner magnetospheric phenomena. Changes in high-latitude

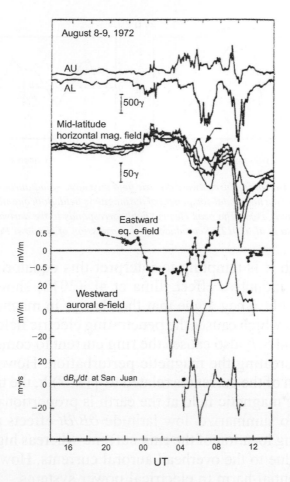

Fig. 5.6 From top to bottom: The AU, AL indices; five superposed mid-latitude magnetograms (Kakioka, Tashkent, Tangerang, San Juan, Honolulu) with the arrow pointing to the San Juan magnetogram; the zonal electric field at the equator (Jicamarca, Peru); the auroral zonal electric field at College, Alaska; and the time derivative of the horizontal component of the magnetic field at San Juan, Puerto Rico. The magnetic field unit γ has been replaced by nT. In the last three panels, the dots indicate local magnetic midnight. After Gonzales et al. (1979). Reproduced with permission of the American Geophysical Union.

electric and magnetic fields create inner magnetospheric electric fields called prompt penetrating electric fields (PPE) (also see Section 5.5). The San Juan magnetogram shows that these are accompanied by significant $\partial B/\partial t$. Figure 5.7 is another example where $\partial B/\partial t$ at the earth's surface in Peru reached 5 nT/min and was correlated with a PPE event.

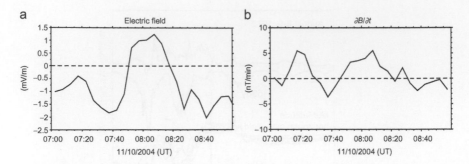

Fig. 5.7 Comparison of (a) the eastward electric field measured simultaneously with (b) the time derivative of the horizontal component of the magnetic field. Both quantities were measured at Jicamarca, Peru. The peak electric field corresponds to the derivative of B near 08:00. After Ilma et al. (2012). Reproduced with permission of Hindawi Publishing.

Although it is tempting to interpret this equatorial electric field as an inductive effect, Ilma et al. (2012) show that this is not credible. They argue that the change in magnetospheric convection, which causes the penetrating electric field shown in Figs. 5.6 and 5.7, also causes the ring current to come closer to the earth, creating the magnetic perturbation. However, since the location of the current is related to $\int (E/B)dt$, the time derivative of the magnetic field at the earth is proportional to δE of the PPE. To summarize, low latitude $\partial B/\partial t$ effects are due to the changing location of the ring current, whereas high latitude $\partial B/\partial t$ are due to the overhead auroral currents. However, both are of potential harm to electrical power systems.

5.2.3 Electric Field Effects on the Atmosphere

Many other processes accompany magnetic storms and substorms. The plasma is only a small fraction of the neutral density but, with time, it can drive the winds to high speed. An example of this interaction is presented in Fig. 5.8. Here, a number of neutral and plasma clouds were released in the auroral zone. The ion (barium) clouds took off with the $\mathbf{E} \times \mathbf{B}$ drift to the west. The neutral TMA and strontium (a small addition to the barium material) clouds followed with up to one-third of the plasma's velocity. The time constant for such

a

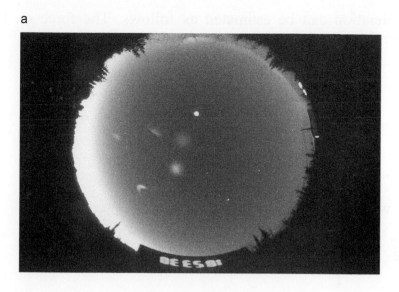

b

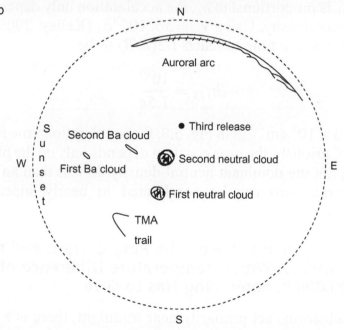

Fig. 5.8 (a) An all-sky camera photo showing the position of ion and neutral clouds in the BaTMAn experiment several minutes after the second barium release. Here, Ba stands for the barium release and TMA to the trimethyl aluminum trails used for neutral wind measurements. The bright dot [see (b)] is the third barium release at the detonation time. (b) Key for identifying features in part (a). After Kelley (2009). Reproduced with permission of Elsevier.

acceleration can be estimated as follows. The force on the neutral gas in the middle ionosphere is

$$\rho \frac{\partial \mathbf{u}}{\partial t} = \mathbf{J} \times \mathbf{B} = \sigma_\rho EB.$$

Using

$$\sigma_\rho = \frac{ne^2 \nu_{in}}{B^2},$$

the wind acceleration is

$$\frac{\partial \mathbf{u}}{\partial t} = \left(\frac{n\nu_{in}}{n_n}\right)\left(\frac{E}{B}\right).$$

Since ν_{in} is proportional to n_n, the acceleration only depends on the plasma density. Using $\nu_{in} = 5 \times 10^{-9} n_n$ (Kelley, 2009), we can estimate the time to reach $1/3(E/B)$ to be

$$\partial t_{1/3} = \frac{10^{10}}{1.5n}.$$

For $n = 3 \times 10^5$ cm^{-3} as in Fig. 5.8, the acceleration time is only 2000 s. Curiously, the time constant depends only on the plasma density, not the dominant neutral density. In less than an hour, the neutrals thus can be accelerated to nearly supersonic velocities.

5.2.4 Interaction Between the Ring Current and the Plasmasphere: With a Temperature Difference of 300,000,000 K, Something Has to Give

Since substorms act primarily near midnight, there is a lot of excess heat along these field lines. This means that the usual radial pressure gradient, which drives the ring current, has a small azimuthal gradient pointing from dusk toward midnight. As we shall see, this has a huge effect on the ionospheric electric field near the boundary. In fact, it has been known for a long time (Smiddy et al., 1977) that large, localized, poleward-directed electric fields develop during high magnetic activity at the equatorward edge of the auroral zone. The potential across this region is typically 25 keV (Rich et al., 1981). The size of the region is anticorrelated with the magnitude of the electric field, creating a nearly identical potential for all the events. The region is located at the inner edge of the diffuse aurora and is associated with a very large $\mathbf{E} \times \mathbf{B}$ drift. Figure 5.9 is a view from 800 km and shows a distinct scalloping of the edge (Lui et al., 1982). This has been attributed to a Kelvin-Helmholtz instability (Kelley, 1986) and successfully simulated (Yamamoto et al., 1994), as shown in Fig. 5.10.

The origin of this field is not yet known. We argue here that since azimuthal pressure gradients cause radially inward currents to flow, these must be closed in the ionosphere or canceled locally by a polarization electric field. If the classical Spitzer conductivity (σ_s) describes the physics, only a very small electric field is needed. But, if there is an anomalously low conductivity, σ^*, due to wave-particle interactions, the electric field can be much larger. Since the field is observed, a σ^* must be operating. Kelley et al. (2012) have shown that all manner of waves occur in this region of the magnetosphere. There is a consistency between the pressure gradient, the electric field, and the field-aligned currents if $\sigma^* = \sigma_s/10^4$, as found with laboratory studies (Chen, 1984). If the magnetospheric conductivity is lower than its classical value, an electric field will form to keep its divergence small. The field-aligned current then will flow along the magnetic field toward the ionospheres near the earth and away from it at larger distances. This is shown schematically in Fig. 5.11, which is drawn in the meridional plane. When the current system closes in the ionosphere, the electric field at the boundary maps to the poleward electric field and $\mathbf{J} \cdot \mathbf{E} > 0$, since the ionosphere acts as an electrical load dissipating energy. In the equatorial plane, $\mathbf{J} \cdot \mathbf{E} < 0$, as required for a source.

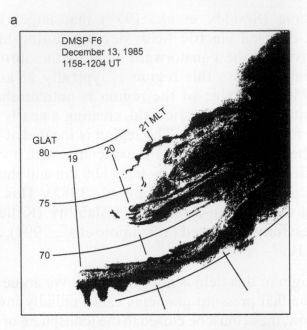

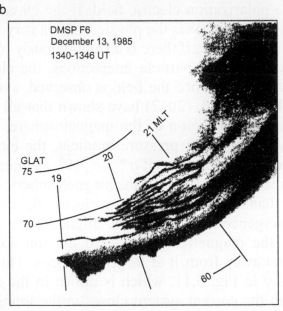

Fig. 5.9 Outlines of auroral images photographed from the DMSP F6 satellite in the Northern Hemisphere at around (a) 12:02 UT and (b) 13:45 UT on December 13, 1985. In (a), giant (type 2) undulations are seen on the equatorward boundary of the diffuse aurora near 19 MLT. The undulations in (b) are thought to be type 2 undulations in the early stages of development. After Yamamoto et al. (1994). Reproduced with permission of the American Geophysical Union.

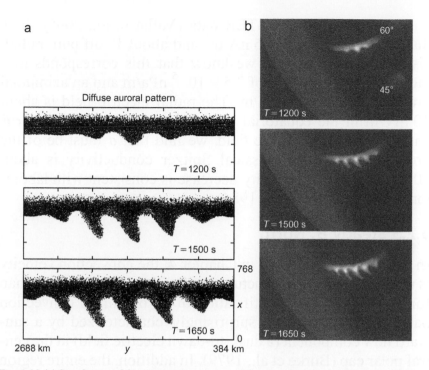

Fig. 5.10 (a) Simulation of the temporal evolution of the distribution of the diffuse aurora on the poleward edge of the trough. (b) A three-dimensional picture of giant undulations with imitative color, obtained by projecting the computer-produced auroral pattern (a) onto the globe surface. After Yamamoto et al. (1994). Reproduced with permission of the American Geophysical Union.

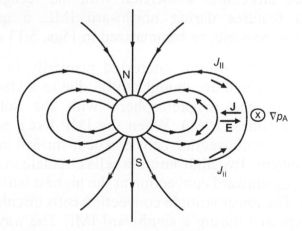

Fig. 5.11 Current systems and the electric field generated at the boundary due to azimuthal pressure gradients, ∇p_A. Adapted from Kelley (2009). Reproduced with permission of Elsevier.

To summarize, CLUSTER data (Vallat et al., 2005) show that the ring current is 55 nA/m^2 and about 1° off pure radial. Using ionospheric data, we know that this corresponds to a radial pressure gradient of 2.5×10^{-5} nPa/m and an azimuthal gradient of 5×10^{-7} nPa/m. The radial electric field is about 40 mV/m in the equatorial plane. Using $\mathbf{J} = \mathbf{B} \times \nabla \mathbf{p}/\mathbf{B}^2 = \sigma \mathbf{E}$ and the observed electric field, we find that σ must be of the order of 0.4 S. The classical Spitzer conductivity is about 1000 S, so the conductivity decrease is consistent with laboratory experiments (Chen, 1984).

5.2.5 Steady B_z North

In this case, as discussed in Chapter 4, the convection velocity is usually much more structured and of smaller magnitude than for B_z south. When identification of an organized convection pattern is possible, it is, surprisingly, characterized by a sunward flow component (a dusk-to-dawn electric field) in the central polar cap (Burke et al., 1979). In addition, the entire region of significant plasma motion is confined to much higher latitudes than in the southward IMF case. Figure 5.12 shows several examples of the dawn-to-dusk electric field measured at high latitudes when the IMF had a northward component. Despite the difficulties associated with the recognition of convection features during northward IMF, a qualitative description is possible, as summarized in Figs. 5.13 and 5.14.

In this chapter, we are interested primarily in the low-latitude cells, which are thought to be due to a viscous-like interaction of the magnetosphere with the solar wind (Axford and Hines, 1961). When the IMF has a northward component, four convection cells can be identified in the dayside hemisphere. Two high-latitude cells circulate in a manner that produces sunward convection at the highest latitudes (see Fig. 5.12). The lower latitude convection cells circulate in the manner expected during a southward IMF. The way viscous interaction can cause this is illustrated in Fig. 5.13. The total potential drop across these "low" latitude cells rarely exceeds 10 kV and it may well be that this part of the convection

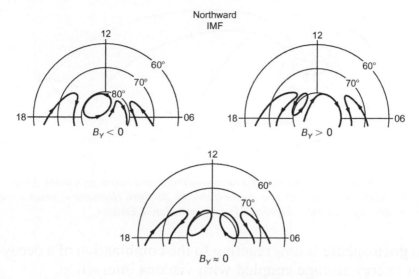

Fig. 5.12 *The main feature of the dayside convection geometry, when the IMF has a north-ward component, is the existence of four convection cells. After Heelis et al. (1986). Reproduced with permission of the American Geophysical Union.*

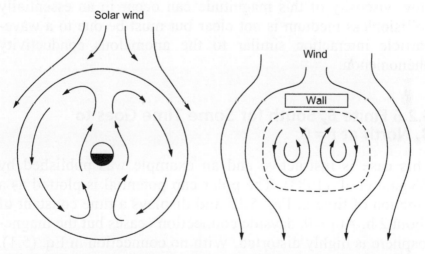

Fig. 5.13 *Schematic diagram showing how a viscous interaction could drive magnetospheric convection, along with a fluid analogy. After Kelley (2009). Figure courtesy of D.P. Stern and S. Lantz and reproduced with permission of Elsevier.*

pattern is driven by a viscous interaction described schematically in Fig. 5.14. It is important to note that outside the region of connected field lines in the deep polar cap, the two-cell pattern has the same shape as the B_z south case. However, this occurs because, for field lines that are closed, the

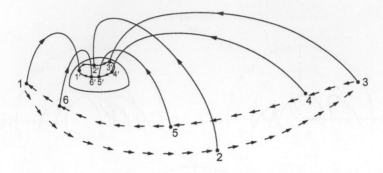

Fig. 5.14 Time history of convecting flux tubes resulting from a viscous interaction. Here, all the flux tubes are closed and also connect into the Southern Hemisphere, which is not shown. After Kelley (2009). Reproduced with permission of Elsevier.

magnetosphere is now reacting to the combination of a decaying energy storage coupled with viscous interaction.

Such a magnetospheric circulation must be accompanied by an electric field, $\mathbf{E} = -\mathbf{V} \times \mathbf{B}$, which maps to the ionosphere. How viscosity of this magnitude can occur in an essentially collisionless medium is not clear but must be due to a wave-particle interaction similar to the anomalous conductivity phenomenon.

5.2.6 Finite B_z South for Some Time Goes to B_z North at $t = 0$

This case is instructive and an example was published by Wygant et al. (1983). The polar cap potential is plotted as a function of time in Fig. 5.15 and displays a time constant of about 2 h. At $t = 0$, dayside connection ceases but the magnetosphere is highly distorted. With no connection in Eq. (5.1), there is no electric field parallel to the magnetopause and $\mathbf{E} \times \mathbf{H} = 0$. The magnetosphere is a high-energy state and the left-hand side must thus be negative. This energy decay comes through the Joule heating and work done represented by the right-hand volume integral. Kelley (2012) has shown that the time constant for this description is consistent with Fig. 5.15.

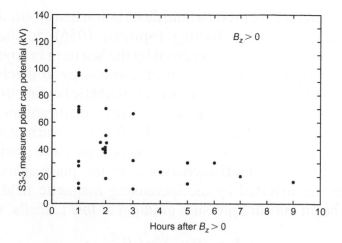

Fig. 5.15 The polar cap potential decreases slowly after the IMF turns northward. It finally reaches a value between 20 and 15 kV that is perhaps attributable to the viscous interaction process. After Wygant et al. (1983). Reproduced with permission of the American Geophysical Union.

5.3 THE SOURCE OF REGION 2 CURRENTS

We will concentrate again on steady B_z south conditions. Then, in the tail, a steady sunward convection occurs in which plasma moves earthward into an ever increasing magnetic field line. As it does so, the plasma becomes hotter and hotter. This can be understood in two ways. From a single particle standpoint, the particles are energized by conservation of their first adiabatic invariant (Kelley, 1989, 2009), $\mu_\perp$, where

$$\mu_\perp = \left(\frac{1}{2} m V_\perp^2 / B_0 \right).$$

In the tail, B_0 might be as small as 4 nT and while at $L=4$, it is 4000 nT. A 1 keV particle is thus energized to 1 MeV as it connects inward. This is the origin of the outer Van Allen radiation belts and a major source of space weather problems. But as particle energy rises, so do the magnetic forces. The so-called gradient and curvature drifts begin to create zonal drifting, which eventually exceeds the $\mathbf{E} \times \mathbf{B}$ drift for the most energetic components (these form the long-lasting outer radiation belts).

In this chapter, we prefer a fluid description that is on a slightly more solid theoretical footing (Spitzer, 1956). In the fluid description, particles are energized by the last term in Poynting's Theorem. This term converts electrical energy to mechanical energy or heat and is analogous to the adiabatic heating in single particle theory. The gradient and curvature drifts are related to the pressure-related forces and the $\mathbf{J} \times \mathbf{B}$ term in the momentum equation. It is natural to expect an inward pressure gradient at high L shells and, as $\mathbf{B}$ increases near the plasmapause and particles are diverted by the increasing magnetic field, there should be an outward pressure gradient at low L shells. Using

$$\mathbf{J}_\perp = \mathbf{B} \times \nabla p / |\mathbf{B}|^2,$$

the ring current is west at high L shells and east for low L shells. Curiously, it is the outer current that dominates the magnetic field at the earth, since $|\mathbf{B}|$ decreases faster than $|\nabla p|$. Clearly, the westward current dominates, making a southward δB at the earth, the so-called main phase of a magnetic storm reaching 400 nT in a "super" magnetic storm and 2000 nT in the Carrington storm.

To first order, the zonal ring current is divergence-free. Consider now the situation in Fig. 5.16 and concentrate on

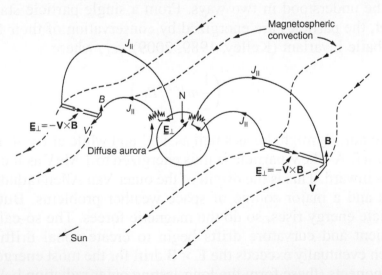

Fig. 5.16 *Three-dimensional view of the electric and magnetic field geometry on auroral zone flux tubes. After Kelley (2009). Reproduced with permission of Elsevier.*

the dawn or dusk sectors. To understand the region 2 currents, we are interested in the direction of the radial current, since that current clearly is not divergence-free and, in fact, is closed in the ionosphere via the parallel currents shown. This region has a small pressure gradient pointing toward the sun. In the dawn sector, $\mathbf{J}_\perp = (\mathbf{B} \times \nabla_p / |\mathbf{B}^2|)$ is radically inward. However, so is the electric field for sunward convection. Thus, $\mathbf{J} \cdot \mathbf{E} > 0$ and no electric generator results. But what, then, drives the radial $\mathbf{J}_\perp$? As with the region 1 currents, it is the slowing down of the plasma as it moves into a high magnetic field. Consider again the momentum equation:

$$\rho \left(\frac{\partial V}{\partial t} + (\mathbf{V} \cdot \nabla \mathbf{V}) \right) = -\nabla p + \mathbf{J} \times \mathbf{B}.$$

This time we need the advective term since, in a steady state, $\partial V / \partial t = 0$. Ignoring ∇p,

$$\mathbf{J}_\perp = \rho (\mathbf{B} \times \mathbf{V} \cdot \nabla \mathbf{V}) / |\mathbf{B}|^2,$$

$$\mathbf{J}_\perp = \frac{\rho}{B} (\hat{B} \times \mathbf{V} \cdot \nabla \mathbf{V}).$$

As the plasma approaches earth, $\mathbf{E} \times \mathbf{B} / |\mathbf{B}|^2$ decreases since $\mathbf{E}$ stays about the same while $|\mathbf{B}|$ increases. Thus, $\nabla \mathbf{V}$ is anti-sunward and $\mathbf{V} \cdot \nabla \mathbf{V}$ is also antisunward. Then $\mathbf{B} \times (\mathbf{V} \cdot \nabla \mathbf{V})$ is outward, $\mathbf{J} \cdot \mathbf{E} < 0$, and we have an MHD generator. This situation is shown in Fig. 5.16. Clearly, $J_{\parallel}$ is closed in the two auroral zones and has the directions shown. The equator-ward pairs feed the region 2 currents and the poleward pairs add a contribution to the region 1 currents. If the hot precipi-tating plasma is copious enough, it will ionize the atmosphere below enough to complete the circuit via the diffuse aurora. Then, most of the squared $\mathbf{J} \cdot \mathbf{E} > 0$ takes place in the E region. But if $J_{\parallel}$ is too high, discrete aurora occurs. As discussed in Section 5.6, such aurora are characterized by parallel electric fields such that $J_{\parallel} \cdot E_{\parallel} > 0$. Thus, energy dissipation occurs at high altitude (≥ 5000 km). But also, the accelerated electrons increase Σ_p in the E region, which also enhances dissipation.

To summarize, region 1 currents (high latitude) partially close in the solar wind and the remainder join the current source shown in Fig. 5.16, closing across the ionosphere and the region 2 current system.

5.4 PENETRATION OF HIGH-LATITUDE ELECTRIC FIELDS TO LOW LATITUDES

As we have seen in Chapter 4, the interplanetary magnetic field direction controls the efficiency of the solar wind MHD generator. In particular, southward turnings of the IMF create the most efficient conditions for magnetic imaging. Long-period variations of the generator and of the magnetospheric generator are shielded by the Alfvén layer discussed earlier in this chapter. However, this system acts like a high-pass filter that is incapable of shielding higher frequency electric fields. A classic example is shown in Fig. 5.17 where the magnetic field in this interplanetary medium is compared with the magnetic field due to the equatorial electrojet as measured in Huancayo, Peru (Nishida, 1968). For the 1-h fluctuations, the two signals are quite similar, even though separated by 105 km. The equatorial

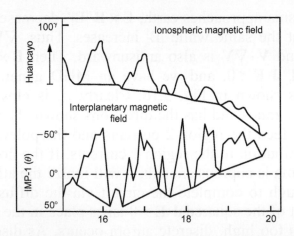

Fig. 5.17 Correlation of horizontal geomagnetic fluctuations observed at Huancayo on December 3, 1963, with changes in the direction of the interplanetary magnetic field component perpendicular to the sun-earth line observed from the satellite IMP-1 versus UT. After Nishida (1968). Reproduced with permission of the American Geophysical Union.

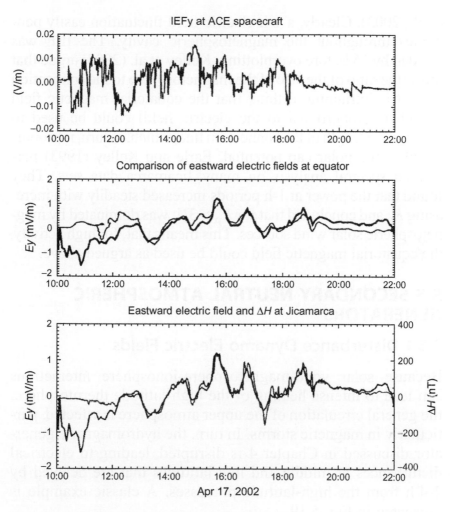

Fig. 5.18 *Direct comparison of the IEF and the equatorial electric field (using ISR and the Anderson technique) after adjusting for the time delay from the ACE satellite to the magnetopause. In the middle and bottom panels, the dark line is the eastward component over Jicamarca that corresponds to the dawn-to-dusk direction. The light line in the middle panel is the dawn-to-dusk component of the IEF. In the lower panel, the light line is the predicted eastward field from ΔH values. After Kelley et al. (2003). Reproduced with permission of the American Geophysical Union.*

electrojet is driven by the eastward electric field component at the magnetic equator. A more direct example is presented in Fig. 5.18 where electric field data from the Jicamarca Radio Observatory are compared to the interplanetary electric field (IEF) measured on the ACE satellite in the solar wind (Kelley et al., 2003). Clearly, a 1-h electric field fluctuation easily penetrates throughout the magnetospheric cavity. The IEF was divided by 15 before overplotting. Kelley et al. (2003) argue that this is the ratio of the size of the magnetosphere to the connection line. It is intriguing to think that the equatorial magnetic field (which is proportional to the electric field) could be used to monitor the length of the connection line, which, in turn, is proportional to the polar cap potential. Earle and Kelley (1993) performed spectral analysis on many Jicamarca data sets. They found that the power at 1-h periods increased steadily with increasing K_p and concluded that for $K_p > 3$, it was dominated by magnetospheric solar wind sources. This means that, for high activity, the equatorial magnetic field could be used as argued above.

5.5 SECONDARY NEUTRAL ATMOSPHERIC GENERATORS

5.5.1 Disturbance Dynamo Electric Fields

Because solar wind-magnetosphere-ionosphere interactions can lead to intense heating of the high-latitude thermosphere, the general circulation of the upper atmosphere is affected, particularly in magnetic storms. In turn, the hydromagnetic generator discussed in Chapter 4 is disrupted, leading to electrical disturbances at middle and low latitudes that are delayed by 2-4 h from the high-latitude processes. A classic example is presented in Fig. 5.19.

In the lower panel, the electric field in Peru slowly reverses sign over a several-hour period, presumably as a wind pulse propagates out of the auroral zone. Notice that the K_p levels have decreased from a peak value many hours before the event. The panel just above is an example of PPE the day before. An identical waveform was seen by the Chatanika radar. Figure 5.20 shows the wave-like nature of such an event as it passes over Arecibo (Nicolls et al., 2004). Notice the downward phase progression characteristic of a gravity wave (Hines, 1974).

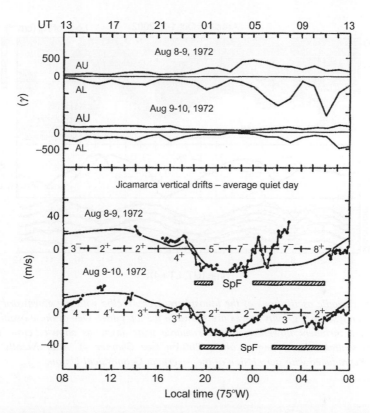

Fig. 5.19 Auroral magnetic fields and F-region vertical drifts at Jicamarca on August 8-10, 1972. The solid curves in the lower panel show the average quiet-time diurnal variation. Deviations from this pattern beginning at 23:00 LT on August 8 are due to direct penetration effects, whereas the slower deviations starting at 22:00 LT on August 9 are due to the disturbance dynamo. After Fejer et al. (1983). Reproduced with permission of the American Geophysical Union.

5.5.2 Flywheel-Generated Electric Fields

At high latitudes, the applied electric fields to first order form a two-celled plasma flow pattern that is virtually always present. It is not surprising, then, that the high-latitude neutral atmospheric motions are greatly affected by electrodynamic forcing. Winds in the thermosphere are driven across the polar cap by the plasma flow. The plasma turns to follow the auroral oval but the wind, to first order, has no such constraint. A flywheel-like effect may then occur with disturbance winds

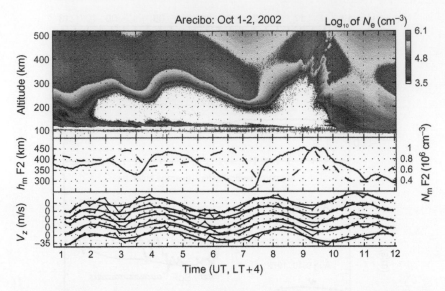

Fig. 5.20 Periodic oscillations of the ionosphere over Arecibo with the downward phase progression typical of gravity waves are shown in the top panel. One of these oscillations triggered a turbulent upwelling over Jicamarca near dawn, as shown below, in the coherent scatter echoes ranging above 200 km (see Chapter 4). After Nicolls et al. (2004). Reproduced with permission of the American Geophysical Union.

blowing out of the auroral oval, even during relatively quiet times. Such winds would be reinforced by the equatorward pressure gradient due to Joule and particle heating in the auroral oval. Once equatorward of the oval, the force might subsequently deflect the wind toward the west. This wind would then create a disturbance dynamo electric field in the poleward direction.

During very active times, there is good evidence for such a wind pattern. Quiet-time neutral wind measurements at Fritz Peak, Colorado (39.9°N, 105.5°W, $L=3$) for six nights are gathered in Fig. 5.21, along with the predictions of a thermospheric global circulation model (Hernandez and Roble, 1984). The data and model both show eastward winds in the evening sector. On the other hand, measurements made on an active day as shown in Fig. 5.21 display strong westward winds until 02:00 LT and an equally strong equatorward

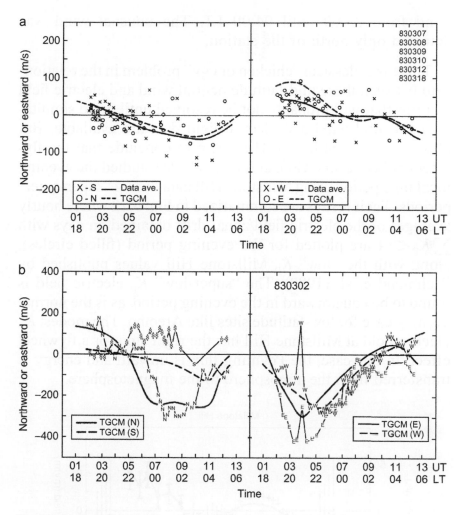

Fig. 5.21 (a) Nighttime variation of thermospheric winds measured during six geomagnetic quiet days in early March 1983 with the year, month, and day given in the upper right corner. The meridional and zonal wind measurements, positive northward and eastward, are given at the left and right, respectively, with the solid line being an average of the data points and the dashed line representing TGCM predictions for geomagnetic quiet conditions. (b) Nighttime thermospheric winds measured on March 2, 1983: (left) meridional winds (positive northward) measured to the north (N) and south (S); (right) zonal winds (positive eastward) measured to the east (E) and west (W) of Fritz Peak Observatory. The solid and dashed curves represent TGCM predictions for constant-pressure surface near 300 km and for grid points north and south (left) and east and west (right) from Fritz Peak. After Hernandez and Roble (1984). Reproduced with permission of the American Geophysical Union.

wind from 23:00 until 04:00 LT. The unusual wind was detected only north of the station.

There is a classical "chicken or egg" problem in the relationship between these mid-latitude neutral wind and electric field patterns that will probably not be resolved until more simultaneous neutral wind and electric field data are available. But observations at Millstone Hill seem to conclude that the flywheel effect occurs. Gonzales et al. (1978) studied the evening local time period using Millstone Hill data during "super-quiet" periods. Their results are summarized in Fig. 5.22 where hourly averages of the electric field/zonal drift obtained on days with $\sum K_\mathrm{p} < 14$ are plotted for the evening period (filled circles), along with the "low" K_p Millstone Hill values published by Richmond et al. (1980). The "super-low" K_p electric field is found to be equatorward in the evening period, as is the normal evening case for low-latitude sites like Arecibo. The modest K_p electric field at Millstone Hill has the right sign for a flywheel effect. In this case, the Poynting flux is upward and energy is transferred from the atmosphere to the magnetosphere.

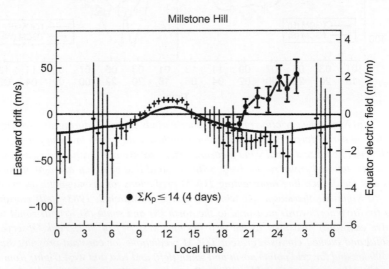

Fig. 5.22 Average Millstone Hill equatorward electric field component measurements at equinox along with 1-h averages of the same component during times of very low activity. After Kelley (2009). Figure courtesy of C. Gonzales and reproduced with permission of Elsevier.

5.6 PARALLEL ELECTRIC FIELDS

One of the most spectacular of earth's natural phenomena is the aurora. A spectacular example is shown in Fig. 5.23 as seen from above from one of the DMSP satellites. This vast expanse of unstructured light is due to precipitation of high plasma stored in the distorted high-latitude magnetic field. This is called the

Fig. 5.23 Defense Meteorological Satellite Program (DMSP) image taken from 800 km over a piece of the auroral oval. Hudson Bay is at the top, Florida and Cuba at bottom right. The bright spots are cities. The dark spot is Ithaca, NY. After Kelley (2002). Figure courtesy of F. Rich and reproduced with permission of Elsevier.

diffuse aurora and extends from the edge of the polar cap to the edge of the ring current. This 10 million K plasma indeed carries the ring current and is the source of the Alfvén layer that shields the plasmasphere from the external electric field at low frequencies.

Some of the earliest rocket experiments were made on "rockoons," small rockets carried aloft by balloons. They found that discrete aurora were characterized by a nearby monoenergetic distribution function rather than an elevated Maxwellian, as found in the diffuse aurora. Very early experimentalists suspected that electric fields developed parallel to the magnetic field, accelerating electrons down and ions up. Theorists had a hard time with this since one of the most cherished axioms is $E_\parallel = 0$ due to the high conductivity parallel to **B**. Magnetic field lines were perfectly conducting wires, period.

This construct broke down in the mid-1970s. First, barium-shaped charge releases from rockets reached altitudes where the ions suddenly took off away from the earth, easily escaping gravity (Wescott et al., 1976). Figure 5.24 shows one example in which a parallel potential drop of 7.4 kV was detected, very similar to the monoenergetic peaks in discrete aurora. The smoking gun had been found. At the same time, satellite S3-3 found huge, perpendicular electric fields above 5000 km, fields never observed in the ionosphere (Mozer et al., 1977). There simply had to be a potential drop below the spacecraft.

Figure 5.25 shows the classes of potential structure seen by S3-3 and later by FAST. Close lines mean high fields, and the spreading at low altitude shows the decreased perpendicular **E**. The parallel **E** lies between about 1 Re and the discrete aurora below. These potential patterns have corresponding electron energy structures below called "inverted V's." Examples are shown in Fig. 5.25 for both regular aurora and black aurora in which the parallel field is reversed (Carlson et al., 1998). As you approach the region, the average energy rises, peaks, and decreases again in an inverted V shape.

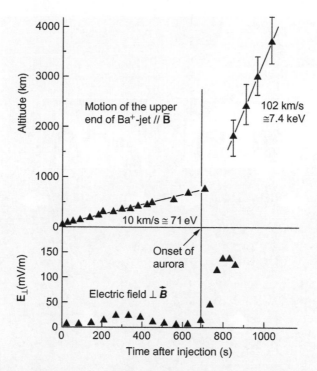

Fig. 5.24 Upward velocity of the upper end of the barium ion jet (January 11, 1975) (upper panel) and transverse electric field projected to the 100 km level as derived from the low altitude portion of the ion jet (lower panel). After Kelley (2009). Reproduced with permission of Elsevier.

Several theories have evolved to explain this phenomenon. The first merely involved an anomalously low parallel conductivity, $\sigma^* = ne^2/mv^*$, where wave-particle interactions created collisions that supported the electric field. When the current becomes too high in this model, ion acoustic waves are generated, causing collisions. The threshold for this instability is:

$$V_{th} = \frac{J_{\parallel}}{ne} > \left(\frac{kT_e}{m_e}\right)^{1/2}. \tag{5.6}$$

This criterion is difficult to attain, even for large $J_{\parallel}$. A lower threshold exists for some cyclotron waves if $T_e \gg T_i$, namely

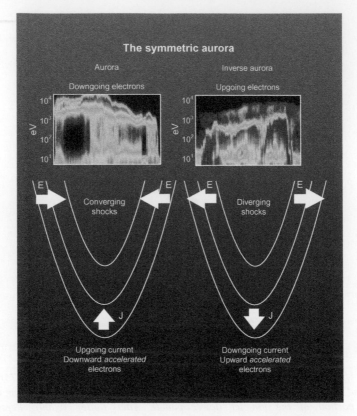

Fig. 5.25 Energetic electrons and ions for opposite types of accelerating structures that have scales of about 3 km. After Kelley (2009). Figure courtesy of C.W. Carlson and reproduced with permission of Elsevier.

$$V_e = \frac{J_\parallel}{ne} = \left(\frac{kT_e}{m_i}\right)^{1/2}. \tag{5.7}$$

But v^* is much lower and the heating is still a problem. Occasional observations of ion cyclotron waves exist.

Many researchers appeal to kinetic or shear Alfvén waves as an explanation. In this approach, the potential structures are waves propagating slowly across the magnetic field. Then, $E_\parallel$ is part of the wave potential. The auroral arc size is then related to the perpendicular wavelength of the shear Alfvén wave.

REFERENCES

Axford, W.I., Hines, C.O., 1961. A unifying theory of high-latitude geophysical phenomena and geomagnetic storms. Can. J. Phys. 39 (10), 1433–1464.

Burke, W.J., Kelley, M.C., Sagalyn, R.C., Smiddy, M., Lai, S.T., 1979. Polar cap electric field structures with a northward interplanetary magnetic field. Geophys. Res. Lett. 6 (1), 21–24.

Carlson, C.W., McFadden, J.P., Ergun, R.E., Temerin, M., Peria, W., Mozer, F.S., Klumpar, D.M., Shelley, E.G., Peterson, W.K., Moebius, E., Elphic, R., Strangeway, R., Cattell, C., Pfaff, R., 1998. FAST observations in the downward auroral current region: energetic upgoing electron beams, parallel potential drops, and ion heating. Geophys. Res. Lett. 25 (12), 2017–2020.

Chen, F.F., 1984. Introduction to Plasma Physics and Controlled Fusion, second ed. Plasma Physics, vol. 1. Plenum Press, New York.

Earle, G.D., Kelley, M.C., 1993. Spectral evidence for stirring scales and two-dimensional turbulence in the auroral ionosphere. J. Geophys. Res. 98 (A7), 11543–11548.

Fejer, B.G., Larsen, M.F., Farley, D.T., 1983. Equatorial disturbance dynamo electric fields. Geophys. Res. Lett. 10 (7), 537–540.

Gonzales, C.A., Kelley, M.C., Carpenter, L.A., Holzworth, H., 1978. Evidence for a magnetospheric effect on mid-latitude electric fields. J. Geophys. Res. 83 (A9), 4397–4399.

Gonzales, C.A., Kelley, M.C., Fejer, B.G., Vickrey, J.F., Woodman, R.F., 1979. Equatorial electric fields during magnetically disturbed conditions, 2. Implications of simultaneous auroral and equatorial measurements. J. Geophys. Res. 84 (A10), 5803–5812.

Heelis, R.A., Reiff, P.H., Winningham, J.D., Hanson, W.B., 1986. Ionospheric convection signatures observed by DE-2 during northward interplanetary magnetic field. J. Geophys. Res. 91 (A5), 5817–5830.

Hernandez, G., Roble, R.G., 1984. Nighttime variation of thermospheric winds and temperatures over Fritz Peak Observatory during the geomagnetic storm of March 2, 1983. J. Geophys. Res. 89 (A10), 9049–9056.

Hines, C.O., 1974. The Upper Atmosphere in Motion: A Selection of Papers with Annotation. Geophysical Monograph 18, American Geophysical Union, Washington, DC.

Ilma, R.R., Kelley, M.C., Gonzales, C.A., 2012. On a correlation between the ionospheric electric field and the time derivative of the magnetic field. Int. J. Geophys. 2012, 648402. http://dx.doi.org/10.1155/2012/648402.

Kappenman, J.G., 2005. An overview of the impulsive geomagnetic field disturbances and power grid impacts associated with the violent Sun–Earth connection events of 29–31 October 2003 and a comparative evaluation with other contemporary storms. Space Weather 3, S08C01.

Kelley, M.C., 1986. Intense shear flow as the origin of large scale undulations of the edge of the diffuse aurora. J. Geophys. Res. 91 (A3), 3225–3230.

Kelley, M.C., 1989. The Earth's Ionosphere: Plasma Physics and Electrodynamics, International Geophysics Series, vol. 43. Academic Press, San Diego, CA.

Kelley, M.C., 2002. Ionosphere. In: Holton, J.R., Pyle, J.A., Curry, J.A. (Eds.), Encyclopedia of Atmospheric Science. Academic Press, London, p. 1022.

Kelley, M.C., 2009. The Earth's Ionosphere: Plasma Physics and Electrodynamics, second ed. International Geophysics Series, vol. 96. Academic Press, Burlington, MA.

Kelley, M.C., 2012. On the relaxation of magnetospheric convection when B_z turns northward. Ann. Geophys. 30, 927–928.

Kelley, M.C., Makela, J.J., Chau, J.L., Nicolls, M.J., 2003. Penetration of the solar wind electric field into the magnetosphere/ionosphere system. Geophys. Res. Lett. 30 (4), 1158. http://dx.doi.org/10.1029/2002GL016321.

Kelley, M.C., Franz, J., Jacobson, A., 2012. On structuring of the plasmapause. Geophys. Res. Lett. 39 (1), L01101. http://dx.doi.org/10.1029/2011GL050048.

Lui, A.T.Y., Meng, C.-I., Ismail, S., 1982. Large amplitude undulations on the equatorward boundary of the diffuse aurora. J. Geophys. Res. 87 (A4), 2385–2400.

Mozer, F.S., 1970. Electric field mapping from the ionosphere to the equatorial plane. Planet. Space Sci. 18 (2), 259–263.

Mozer, F.S., 1973. Electric fields and plasma convection in the plasmasphere. Rev. Geophys. 11 (3), 755–765.

Mozer, F.S., Carlson, C.W., Hudson, M.K., Torbert, R.B., Parady, B., Yatteau, J., Kelley, M.C., 1977. Observations of paired electrostatic shocks in the polar magnetosphere. Phys. Rev. Lett. 38 (6), 292–295.

Nicolls, M.J., Kelley, M.C., Coster, A.J., González, S.A., Makela, J.J., 2004. Imaging the structure of a large-scale TID using ISR and TEC data. Geophys. Res. Lett. 31, L09812.

Nishida, A., 1968. Coherence of geomagnetic *DP* 2 fluctuations with interplanetary magnetic variations. J. Geophys. Res. 73 (17), 5549–5559. http://dx.doi.org/10.1029/JA073i017p05549.

Rich, F.J., Cattell, C.A., Kelley, K.C., Burke, W.J., 1981. Simultaneous observations of auroral zone electrodynamics from two satellites: evidence for height variations in the topside ionosphere. J. Geophys. Res. 86 (A11), 8929–8940.

Richmond, A.D., Blanc, M., Emery, B.A., Wand, R.H., Fejer, B.G., Woodman, R.F., Ganguly, S., Amayenc, P., Behnke, R.A., Calderon, C.,

Evans, J.V., 1980. An empirical model of quiet-day ionospheric electric fields at middle and low latitudes. J. Geophys. Res. 85 (A9), 4658–4664.

Smiddy, M., Kelley, M.C., Burke, W., Rich, R., Sagalyn, R., Shuman, B., Hays, R., Lai, S., 1977. Intense poleward-directed electric fields near the ionospheric projection of the plasmapause. Geophys. Res. Lett. 4 (11), 543–546.

Spitzer Jr., L., 1956. Physics of Fully Ionized Gases. Interscience Publishers, Inc., New York.

Vallat, C., Dandouras, I., Dunlop, M., Balogh, A., Lucek, E., Parks, G.K., Wilber, M., Roelof, E.C., Chanteur, G., Rème, H., 2005. First current density measurements in the ring current region using simultaneous multi-spacecraft CLUSTER-FGM data. Ann. Geophys. 23 (5), 1849–1865.

Wescott, E.M., Stenbaek-Nielsen, H.C., Hallinan, T.J., Davis, T.N., Peek, H.M., 1976. The Skylab barium plasma injection experiments, 2. Evidence for a double layer. J. Geophys. Res. 81 (A25), 4495–4502.

Wygant, J.R., Torbert, R.B., Mozer, F.S., 1983. Comparison of S3-3 polar cap potential drops with the interplanetary magnetic field and models of magnetopause reconnection. J. Geophys. Res. 88 (A7), 5727–5735.

Yamamoto, T., Ozaki, M., Inoue, S., Makita, K., Meng, C.-I., 1994. Convective generation of "giant" undulations on the evening diffuse auroral boundary. J. Geophys. Res. 99 (A10), 19499–19512.

CHAPTER 6

Wave-Related Electric Fields

The Earth's Electric Field. http://dx.doi.org/10.1016/B978-0-12-397886-8.00006-5   149

Electrostatic and electromagnetic waves play an important role in the earth's environs. Much of what we call space weather is due to plasma instabilities in the ionosphere. In addition, we use radio waves of many types to probe and understand the system. In this chapter, we derive the key equations of electromagnetic wave theory and apply them to several examples, including lightning. We also explore a few important electrostatic instabilities.

6.1 THE WAVE EQUATION

We start with three of Maxwell's equations for an isotropic medium with permittivity ε and $\mu = \mu_0$:

$$\nabla \cdot E = \frac{\rho}{\varepsilon}, \tag{6.1}$$

$$\nabla \times \mathbf{B} = \mu_0 + \mu_0 \varepsilon \frac{\partial E}{\partial t}, \tag{6.2}$$

$$\nabla \times \mathbf{E} = -\frac{\partial \mathbf{B}}{\partial t}. \tag{6.3}$$

Ignoring sources of current and charge, $J = \rho = 0$. Then, taking the curl of (Eq. 6.3), we have,

$$\nabla \times (\nabla \times \mathbf{E}) = -\frac{\partial(\nabla \times \mathbf{B})}{\partial t} = -\mu_0 \varepsilon \frac{\partial^2 E}{\partial t^2}.$$

Using the vector identity,

$$\nabla \times (\nabla \times \mathbf{A}) = \nabla(\nabla \times \mathbf{A}) - \nabla^2 \mathbf{A},$$

and

$$\nabla \cdot \mathbf{E} = 0,$$

$$\nabla^2 \mathbf{E} - \mu_0 \varepsilon \frac{\partial^2 \mathbf{E}}{\partial t^2} = 0.$$

This is the wave equation and has plane wave solutions for propagation in the z direction of the form

$$\mathbf{E} = \mathrm{Re}\left\{E_1 \widehat{a}_x e^{i(\omega t - kz)}\right\}$$

or

$$\mathbf{E} = \mathrm{Re}\left\{E_2 \widehat{a}_y e^{i(\omega t - kz)}\right\}.$$

With ε a scalar and ω real, $k = 2\pi/\lambda$ is determined by

$$k^2 = \omega^2 \mu_0 \varepsilon$$

and we have

$$k = \omega \sqrt{\mu_0 \varepsilon}$$

where the speed of light, c, equals $1/\sqrt{\mu_0 \varepsilon_0} = 3 \times 10^8$ m/s in free space. For a dielectric medium such as glass, $\varepsilon > \varepsilon_0$ and the speed of an electromagnetic wave is less than c by the factor $(\varepsilon_0/\varepsilon)^{1/2}$. In more complicated media, k may be a complex number and hence may have real and imaginary parts if ε is complex.

In a plasma, it is easiest to explore different frequency regimes separately. In each case, we determine the current carried by ions and electrons and then couch them in the form of a dielectric. In general, for a magnetized plasma, ε is a tensor of the form

$$\widetilde{\varepsilon} = \begin{pmatrix} \varepsilon_{11} & \varepsilon_{12} & 0 \\ \varepsilon_{21} & \varepsilon_{22} & 0 \\ 0 & 0 & \varepsilon_{33} \end{pmatrix}$$

with $\varepsilon_{11} = \varepsilon_{22}$ and $\varepsilon_{21} = \varepsilon_{12}^*$ where $*$ implies a complex conjugate.

Although we have used the concept of a permittivity, plasmas are not generally characterized by polarized molecules as in a dielectric material such as glass. In fact, it is the particle currents that matter. To reconcile these disparate concepts, consider the Maxwell equation:

$$\nabla \times \mathbf{H} = \sigma \mathbf{E}$$

which, for harmonic behavior becomes

$$-i\mathbf{k} \times \mathbf{H} = \sigma \mathbf{E}.$$

If we define an ε_{eff} by

$$\varepsilon_{\text{eff}} = \frac{\sigma}{i\omega}$$

then

$$-i\mathbf{k} \times \mathbf{H} = i\omega\varepsilon_{\text{eff}}\mathbf{E}$$

and we thus have converted conductivity into "permittivity." We now investigate electromagnetic waves, beginning at the lowest frequencies.

6.2 ELECTROMAGNETIC WAVES

6.2.1 Waves with $\omega < \Omega_i$

This is the Alfvén wave regime. In general, for a near collisionless magnetized plasma, particles drift in the presence of a force F with the velocity

$$\mathbf{V_D} = \frac{\mathbf{F} \times \mathbf{B}}{eB^2}.$$

If $F = e\mathbf{E}$, the velocity is the same for electrons and ions and no current flows parallel to $\mathbf{E}$. But as $\mathbf{E}$ changes with time, so does $\mathbf{V_D}$, and there is an inertial force given by

$$\mathbf{F} = -M\frac{\partial \mathbf{V_D}}{\partial t}$$

where M is ion mass. Then, using the general formula above,

$$\mathbf{V_D} = -\frac{M(\partial \mathbf{E}/\partial t \times \mathbf{B}) \times \mathbf{B}}{eB^4} = \frac{M}{eB^2}\frac{\partial \mathbf{E}}{\partial t}.$$

So,

$$\mathbf{J} = ne\mathbf{V_D} = \frac{nM}{B^2}\frac{\partial \mathbf{E}}{\partial t}.$$

For this case,

$$\mathbf{J} = \varepsilon\frac{\partial \mathbf{E}}{\partial t} + \varepsilon_0\frac{\partial \mathbf{E}}{\partial t} = \text{particle current} + \text{displacement current}.$$

We now can define an effective permittivity:

$$\varepsilon_{\text{eff}} = \varepsilon_A = \frac{nM}{B^2} + \varepsilon_0 = \varepsilon_0\left(1 + \frac{nM}{B^2\varepsilon_0}\right)$$

where the A stands for Alfvén and the unity factor, 1, is usually ignored. Then,

$$\widetilde{\widetilde{\varepsilon}} = \begin{pmatrix} \varepsilon_A & 0 & 0 \\ 0 & \varepsilon_A & 0 \\ 0 & 0 & \varepsilon_0 \end{pmatrix}$$

for propagation parallel to $\mathbf{B}$,

$$k^2 = \frac{\omega^2\mu_0 nM}{B^2},$$

and

$$V_A = \frac{\omega}{k} = \sqrt{\frac{B^2}{\mu_0\rho}}$$

where $\rho = nM$ is the mass density. This is the Alfvén speed. In the solar wind, V_A is only 100 km/s, meaning that the solar wind is not only supersonic but super-Alfvénic. Alfvén waves thus cannot "tell" the solar wind that it is approaching the magnetosphere, and a shock wave must occur. Nearing the ionosphere, the Alfvén speed rises to about 1000 km/s.

Most changes in plasmas are communicated along magnetic field lines by Alfvén waves. Examples include pressure variations in the solar wind that create waves inside the magnetosphere and can be detected at the surface of the planet; auroral electric fields transmitted to the ionosphere; and even the space shuttle, which generates Alfvén waves.

In any electromagnetic wave,

$$\mathbf{H} = \frac{\mathbf{E}}{\eta}$$

where η is the intrinsic impedance and, for this case,

$$\eta = \frac{\sqrt{\mu_0}}{\varepsilon} = \frac{\sqrt{B^2}}{\rho}.$$

This is the order of 1 Ω in the ionosphere, which is the same order as the ionospheric field-line-integrated conductivity. Applying transmission line theory, this means that, in general, Alfvén waves usually are at least partially reflected from the ionosphere when generated in the magnetosphere or solar wind.

6.3 WAVES WITH $\Omega_i < \omega < \Omega_e$

This is the whistler mode regime. Such waves are generated, among other things, by lightning and, due to their dispersion, result in eerie, audio frequency radio waves that change frequency with time. These waves were first detected on long-distance phone cables, which acted as antennas for these long wavelength waves. The ions are motionless at these high frequencies, whereas the electrons simply $\mathbf{E} \times \mathbf{B}$ drift, leading to a current as the ions are left behind. For propagation parallel to $\mathbf{B}$, transverse waves drive a perpendicular current that leads to the appropriate ε_{eff} in the tensor permittivity:

$$\mathbf{J}_\perp = \frac{-ne(\mathbf{E} \times \mathbf{B})}{B^2}.$$

To solve the wave equation, and anticipating circularly polarized solutions, we create two orthogonal complex velocities: $(v_x + iv_y)e^{i(kz-\omega t)}$ and $(v_x - iv_y)e^{i(kz-\omega t)}$.

For $\Omega_i < \omega < \omega_p$ or Ω_e, we "guess" that, for $\mathbf{k}//\mathbf{B}$, the waves are circularly polarized. Then we take

$$\mathbf{E}_\pm = \text{Re}(E_0\hat{x} \mp i\hat{y})e^{i(\omega t - kz)}$$

where the minus sign is right-hand polarized and the plus sign is left-hand polarized. The epsilon tensor is:

$$\overset{\approx}{\varepsilon} = \begin{pmatrix} \varepsilon_{11} & \varepsilon_{12} & 0 \\ \varepsilon_{21} & \varepsilon_{22} & 0 \\ 0 & 0 & \varepsilon_{33} \end{pmatrix} \begin{bmatrix} E_x \\ E_y \\ 0 \end{bmatrix},$$

$$\nabla \times \mathbf{E} = -i\omega\mu_0\mathbf{H} \quad \text{and} \quad \nabla \times \mathbf{H} = i\omega\varepsilon\mathbf{E},$$

$$\nabla \times \mathbf{E} = \nabla^2\mathbf{E} - \nabla(\nabla \times \mathbf{E}) = \omega^2\mu_0\overset{\approx}{\varepsilon}\ \mathbf{E}.$$

Since $\nabla \times \mathbf{E} = 0$ in free space, we have

$$\frac{d^2}{dz^2}\begin{bmatrix} E_x \\ E_y \\ 0 \end{bmatrix} = \omega^2\mu_0 \begin{pmatrix} \varepsilon_{11} & \varepsilon_{12} & 0 \\ \varepsilon_{21} & \varepsilon_{22} & 0 \\ 0 & 0 & \varepsilon_{33} \end{pmatrix} \begin{bmatrix} E_x \\ E_y \\ 0 \end{bmatrix},$$

using $E_z = 0$ for a uniform plane wave.

We now need to find the ε tensor from the electron equation of motion:

$$\frac{dV_e}{dt} = \frac{q_e}{m}[\mathbf{E} + \mathbf{V} \times \mathbf{B}_0],$$

using $\mathbf{B}_0 \gg \delta\mathbf{B}$ and $\mathbf{E} = \delta\mathbf{E}$ of the wave. For no bulk electron velocity, $d/dt \rightarrow \partial/\partial t$ and

$$j\omega v_{\rm e} = \frac{q_{\rm e}}{m}\left[\mathbf{E} + \mathbf{v}_{\rm e} \times \mathbf{B}_0\right].$$

Breaking into components,

$$j\omega v_{ex} = \frac{q_{\rm e}}{m}E_x + \frac{q_{\rm e}}{m}B_0 v_{ey},$$

$$j\omega v_{ey} = \frac{q_{\rm e}}{m}E_y - \frac{q_{\rm e}}{m}B_0 v_{ex},$$

$$j\omega v_{ez} = \frac{q_{\rm e}}{m}E_z.$$

The first two equations can be solved simultaneously to yield

$$v_{ex} = -i\omega(q_{\rm e}/m)E_x + (q_{\rm e}/m)\Omega_{\rm e}E_y,$$

$$v_{ey} = -(q_{\rm e}/m)\Omega_{\rm e}E_x - j\omega(q_{\rm e}/m)E_y,$$

$$v_{ez} = i(q_{\rm e}/m)E_z,$$

using $\Omega_{\rm e}E_x = -(q_{\rm e}/m)B_0$. Now to find $\widetilde{\widetilde{\varepsilon}}$, we use

$$i\omega\,\widetilde{\widetilde{\varepsilon}}\,\mathbf{E} = i\omega\varepsilon_0\mathbf{E} + \mathbf{J}.$$

Using $\mathbf{J} = nq_{\rm e}\mathbf{v}_{\rm e}$ and

$$\widetilde{\widetilde{\varepsilon}} = i\omega\begin{pmatrix} \varepsilon_{11} & \varepsilon_{12} & 0 \\ \varepsilon_{21} & \varepsilon_{22} & 0 \\ 0 & 0 & \varepsilon_{33} \end{pmatrix}\begin{bmatrix} E_x \\ E_y \\ E_z \end{bmatrix},$$

we find

$$\varepsilon_{11} = \varepsilon_{22} - \varepsilon_0\left[1 + \omega_{\rm p}^2/\Omega_{\rm e} - \omega^2\right],$$

$$\varepsilon_{12} = \varepsilon_{21}^* = \frac{j\omega_{\rm p}^2(\Omega_{\rm e}/\omega)\varepsilon_0}{\Omega_{\rm e}^2 - \omega^2},$$

$$\varepsilon_{33} = \varepsilon_0\left[1 - \omega_{\rm p}^2/\omega^2\right].$$

Returning to the wave equation,

$$-k^2 E_0 = (\hat{x} \mp i\hat{y}) + \omega^2 \mu_0 (\varepsilon_{11} E_x + \varepsilon_{12} E_y) = 0,$$

and

$$k = \omega \sqrt{\mu_0} (\varepsilon_{11} - i\varepsilon_{12}),$$

$$k_{L/R} = \omega \sqrt{\mu_0} \left(1 \pm \frac{\omega_p^2/\omega}{\omega_e \mp \omega} \right),$$

$$k_R = \omega \sqrt{\mu_0 \varepsilon_0} \left(1 + \frac{\omega_p^2/\omega}{\Omega - \omega} \right)^{1/2}.$$

This mode is very important and for $\omega < \Omega_e < \omega_p$, it is called the whistler mode wave. To a good approximation, then,

$$k_R = \frac{1}{c} \left(1 + \frac{\omega/\omega_p^2}{\Omega} \right)^{1/2}.$$

The phase velocity,

$$\frac{\omega}{k_R} = \frac{(\omega \Omega_e)^{1/2}}{\omega_p} c,$$

thus increases as $\omega^{1/2}$. The group velocity is found from

$$k^2 = \frac{1}{c^2} \left(\omega_p^2 \omega / \Omega_e \right)^{1/2}.$$

Taking the derivative,

$$2k \, dh = \left(\omega_p^2 / c^2 \Omega_e \right) d\omega.$$

So,

$$\frac{d\omega}{dk} = \left(2\Omega_e / \omega_p^2 \right) c^2,$$

and

$$V_g = 2\left(\Omega_e \omega / \omega_p^2\right) c.$$

Since the group velocity increases with increasing frequency, the highest frequencies arrive first from an impulsive source like lightning. While listening on earphones to a receiver output, the source decreases with time and a whistler is born.

Figure 6.1 shows a whistler detected on a rocket above a thunderstorm. A frequency of 75 kHz arrives first, with higher

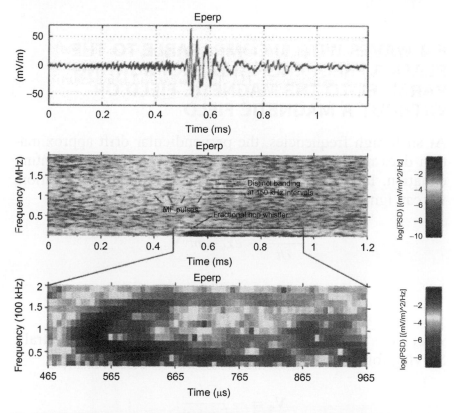

Fig. 6.1 Electric field data taken at 195 km with a 20-kHz to 2-MHz bandwidth receiver. A nose whistler is detected by both sensors at times between 0.5 and 0.7 ms and at frequencies below 175 kHz. A deadband exists between 175 kHz and 1 MHz, which essentially has no electromagnetic energy. Above the deadband are broadband bursts of energy, both before and after the nose frequency arrived. (For a color version of this figure, the reader is referred to the online version of this chapter.) After Kelley et al. (1997). Reproduced with permission of the American Geophysical Union.

and lower frequencies coming later. The above theory (Kelley et al., 1997) is only valid for the lowest frequencies, but a complete analysis shows that a "nose frequency" corresponds to the maximum group velocity. Once whistlers get into the ionospheric/magnetospheric system, they can bounce from one hemisphere to another many times, guided by whistler ducts (plasma waveguides). An early system to measure radial electric fields in the plasmasphere was to watch the progression of these ducts toward or away from the earth, using the nose frequency, which varies with the strength of the magnetic field.

6.4 WAVES WITH ω COMPARABLE TO THE PLASMA FREQUENCY AND PROPAGATING PARALLEL TO THE MAGNETIC FIELD OR WITHOUT A MAGNETIC FIELD

At such high frequencies, the perpendicular drift approximation does not hold and we must use the electron momentum equation. Ions can be taken at rest. For parallel propagation, we can ignore the magnetic field:

$$\frac{m\partial \mathbf{V}_e}{\partial t} - e\mathbf{E} - mv\mathbf{V}_e,$$

$$i\omega \mathbf{V}_e = -\frac{\rho \mathbf{E}}{m} - v\mathbf{V}_e.$$

where v is the collision frequency of electrons and neutrals. Solving for the electron velocity gives

$$\mathbf{V}_e = -\frac{e\mathbf{E}/m}{i\omega + v},$$

and the current is

$$\mathbf{J} = -ne\mathbf{E}_e = \frac{-ne^2\mathbf{E}/m}{i\omega + v}.$$

Table 6.1 Ionosphere summary		
Height (km)	Day	Night
200-400	Collisionless (σ pure imaginary) waves either propagate or reflect	Collisionless (σ imaginary) waves either propagate or reflect
60-200	Collisionless (σ complex) waves attenuate	Dielectric ($\sigma \cong 0$) waves propagate
0-60	Dielectric ($\sigma = 0$) waves propagate	Dielectric ($\sigma = 0$) waves propagate

For this model,

$$\mathbf{J} = \sigma \mathbf{E} + i\omega\varepsilon\mathbf{E},$$

where σ is due to currents carried by particles and ε is due to polarization of molecules, which we ignore.

We thus can turn this into an effective ε by the following:

$$\varepsilon = \frac{\sigma}{i\omega} = \frac{ne}{m(\omega^2 - i\omega v)}. \tag{6.4}$$

The possible cases are summarized in Table 6.1. This formalism works for unmagnetized plasma and for propagation parallel to the magnetic field.

6.4.1 Collisionless Plasmas (σ Imaginary, $\varepsilon = \varepsilon_0$)

During daytime, 0-60 km, and during nighttime, 0-200 km, Table 6.1 has ε real and σ pure imaginary. Such a case can arise if, in Eq. (6.4), we have $\omega \gg v$. Then

$$\sigma = \frac{ne^2}{i\omega m} = -i\left(\frac{ne^2}{\omega m}\right)$$

is purely imaginary. This relation can (and does!) hold in tenuous ionized materials called collisionless plasmas, such as the earth's upper ionosphere. Let $\varepsilon = \varepsilon' - i\varepsilon''$. Then, considering an atomic gas background with $\varepsilon' = \varepsilon_0$ yields

$$k^2 = -\omega^2 \mu\varepsilon_0[1 - \sigma/\omega\varepsilon_0],$$

$$k^2 = -\omega^2 \mu_0 \varepsilon_0 \left[1 - \frac{ne^2}{\omega^2 m \varepsilon_0} \right].$$

The term $ne^2/m\varepsilon_0$ has the units of frequency squared and is called the (radian) plasma frequency:

$$\omega_p^2 = \frac{ne^2}{m \varepsilon_0}.$$

Plugging in the numbers and converting to Hertz, the plasma frequency in Hertz is given by

$$f_p = 8.9\sqrt{n} \text{ Hz},$$

where n is in m^{-3}. Returning to $k^2 = \omega^2 \mu_0 \varepsilon$,

$$k^2 = -\omega^2 \mu_0 \varepsilon_0 \left[1 - \frac{\omega_p^2}{\omega^2} \right],$$

so, whether the wave frequency is larger or smaller than the local plasma frequency is really crucial. If $\omega_p^2/\omega^2 < 1$, then the bracket is positive and, happily, k^2 is positive. This is a propagating solution of the form

$$k = \omega\sqrt{\mu_0 \varepsilon_0} \left(1 - \frac{\omega_p^2}{\omega^2} \right)^{1/2}, \qquad (6.5)$$

where k is real. It is interesting to calculate the phase velocity for this wave:

$$\frac{\omega}{k} = \frac{c}{\left(1 - \dfrac{\omega_p^2}{\omega^2} \right)^{1/2}} > c.$$

The phase velocity exceeds the speed of light, which seems to violate relativity theory. But remember, information travels at the group velocity $d\omega/dk_z$. First, we write $c^2 k^2 = \omega^2 - \omega_p^2$ and note that $2c^2 k dk = 2\omega d\omega$. Hence, $d\omega/dk = c^2/(\omega/k)$. Thus,

$$\frac{d\omega}{dk} = c\left(1 - \frac{\omega_p^2}{\omega^2}\right)^{1/2} < c,$$

so, the relativity theory is safe!

What happens when $\omega_p^2/\omega^2 > 1$? This is something new. The wave number is now imaginary and the phaser is of the form $e^{-\gamma z}$ with no real part of k at all. This *differs* from the skin depth effect that you studied in classes. The skin depth effect had a real and an imaginary part to k. There is no propagation or dissipation here. How can this be?

Imagine a wave propagating into a plasma (like a wave launched toward the ionosphere from the earth) and assume that $\omega < \omega_{p_0}$, the peak plasma frequency at that time and place. In the atmosphere below the ionosphere, $\omega > \omega_{p_0}$ and the wave propagates with no problem. As soon as $\omega < \omega_p$ (as it must at some height if $\omega < \omega_{p_0}$), it cannot propagate because

$$k = \left(i\omega^2\mu\varepsilon_0\left[1 - \frac{i\sigma}{\omega\varepsilon_0}\right]\right)^{1/2}$$

is purely imaginary. Since there is no real part to σ, the power loss is

$$\frac{1}{2}\text{Re}\{\mathbf{J}\cdot\mathbf{E}^*\} = \frac{1}{2}\text{Re}\{\sigma\mathbf{E}\cdot\mathbf{E}^*\} = \frac{|\mathbf{E}|^2}{2}\text{Re}\left\{\sigma = -i\frac{ne^2}{\omega m}\right\} = 0,$$

where Re denotes "take the real part" and there is *no* dissipation. However, conservation of energy requires that the wave energy go somewhere, so it must be perfectly reflected. A medium of this type is often called *reactive* since the conductivity (impedance) is imaginary (reactive).

Figure 6.2 shows a typical ionospheric profile with a peak plasma density of 1012 m^{-3} at about 250 km altitude and the corresponding plasma frequencies. Thus, all waves with $f < 9$ MHz launched from the earth will be reflected back. A

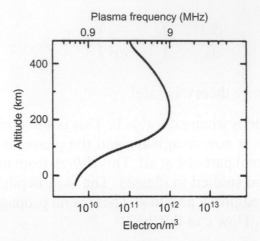

Fig. 6.2 Typical daytime ionospheric plasma density and plasma frequency profiles.

more exact model would have magnetic field effects as well, allowing more wave modes, but for radio, TV, and satellite frequencies, this description is pretty complete. AM radio waves are in the 0.5-1.6 MHz band and bounce off the nighttime ionosphere. This is why listeners on the east coast can easily pick up stations quite far to the west and it explains the way HAM radio works. At a slant angle, you can even skip higher frequencies off the ionosphere. TV and FM radio are near 100 MHz and pass right through, except for very unusual conditions that occur during magnetic or ionospheric storms.

Since *all* satellites are immersed in plasma, it is important for their transmitting system to use frequencies above the highest plasma frequency or the signals simply will not propagate to the earth. However, a reentering vehicle makes its own *very* dense plasma, which is why astronauts have a scary period with no communications when radio waves from the vehicle or the ground cannot penetrate the cloud.

6.4.2 Collisional Plasma (σ = Real and Imaginary, $\varepsilon = \varepsilon_0$)

A medium corresponding to daytime (60-200 km) in Table 6.1 occurs in the lower ionosphere where the atmosphere is still fairly dense. We now model the plasma as a conducting material, but allowing for collisions between electrons and ions with neutrals, Eq. (6.4) still applies.

We now model a plasma as an ionized gas, the $+$ ions having negligible motion as before, but allowing for collisions between electrons and ions or neutrals. The equation of motion for an electron is the same as that for a metal, except the collision frequency, v, is no longer much greater than ω of a wave because the overall density of a plasma is so small:

$$\sigma = \frac{-ne^2}{m(i\omega + v)},$$

but we cannot ignore v now. Rationalizing the denominator:

$$\sigma = \frac{ne^2}{m(\omega^2 + v^2)}(v - i\omega).$$

From the formula for $k^2 = -\omega^2 \mu_0 \varepsilon = -\omega^2 \mu_0 \left(\varepsilon_0 - \frac{i\sigma}{\omega} \right)$,

$$k^2 = -\omega^2 \mu_0 \varepsilon_0 + \frac{ne^2 \mu_0}{m(\omega^2 + v^2)}[\omega^2 + i\omega v],$$

which can be written conveniently as follows:

$$k^2 = -\omega^2 \mu_0 \varepsilon_0 \left(1 - \frac{\omega_p^2}{(\omega^2 + v^2)} - \frac{iv\omega_p^2}{m(\omega^2 + v^2)} \right).$$

For collisionless plasmas, $v = 0$ and we obtain the previous result. For the case where $\omega \gg \omega_p \gg v$, we can simplify the result. Applying this relation in the denominators yields

$$k^2 = -\omega^2 \mu_0 \varepsilon_0 \left(1 - \frac{\omega_p^2}{\omega^2} - \frac{iv\omega_p^2}{\omega^3} \right).$$

Now we let

$$k'' = \omega^2 \mu_0 \varepsilon_0 \left(1 - \frac{\omega_p^2}{\omega^2} \right)$$

and

$$k'' = -\left(k_z' \right)^2 \left(1 - \frac{iv\omega_p^2}{\omega^3 \left(1 - \frac{\omega_p^2}{\omega^2} \right)} \right)$$

$$k'' = -k' \left(1 - \frac{iv\omega_p^2}{\omega^3 \left(1 - \frac{\omega_p^2}{\omega^2} \right)} \right)^{1/2}.$$

If we take $v\omega_p^2 \ll \omega^3$ (consistent with the above assumption), we can use the Taylor series to expand the square root by:

$$(1 - x)^{1/2} \approx 1 - \frac{x}{2},$$

$$k'' = ik \left(1 - \frac{iv\omega_p^2}{2\omega^3 \left(1 - \frac{\omega_p^2}{\omega^2} \right)} \right).$$

We now can identify the real and imaginary parts of k as:

$$k' = \omega^2 \sqrt{\mu_0 \varepsilon_0} \left(1 - \frac{\omega_p^2}{\omega^2} \right)^{1/2},$$

$$k_z'' = \frac{k' v \omega_p^2}{2\omega^3 \left(1 - \frac{\omega_p^2}{\omega^2} \right)}.$$

Thus, a wave propagating in the ionosphere will attenuate due to the real part of σ. Looking back, we see that the assumption of some collisions ($v \neq 0$) is precisely what gave a real part to the conductivity; this is the mechanism for energy loss and wave attenuation in our model. The real part is essentially the same for a collisionless plasma with $\omega < \omega_p$.

We have a real dissipation effect and energy is actually left behind in the material, which explains why AM radio does not propagate well during the day. When the sun is shining, it produces electrons quite low in the atmosphere, well below the nighttime ionosphere, as seen in the figure below. Since the atmosphere is dense at these heights, the electrons collide with atoms and dissipate energy. If no plasma is present at low altitudes, the air acts like a perfect dielectric and the wave passes through to reflect off the high-altitude, collisionless (no dissipation) nighttime ionosphere. The most severe effects occur when $\omega \approx v$, for which the full equation k'' is necessary. The algebra required to study the general case is far beyond this treatise. We refer the reader to Stix (1962, 1992).

6.5 ELECTROSTATIC WAVES IN THE F REGION

In the F region of the ionosphere, electrostatic waves have low phase velocities, the most common of which are called flute modes and can be driven by several types of current perpendicular to the magnetic field. These phenomena are summarized in Table 6.2, where v_{in} is the ion-neutral collision frequency.

Table 6.2 Ionosphere summary for the F region			
Force	Current	Name	Growth ratio
E	$\dfrac{ne^2 v_{in}}{M\Omega_i^2} E$	$\mathbf{E} \times \mathbf{B}$	$\dfrac{E}{BL}$
g	$\dfrac{neg}{v_{in}}$	Rayleigh-Taylor	$\dfrac{g}{v_{in}L}$
$-Mv_{in}U$	$\dfrac{ne^2 v_{in} UB}{M\Omega_i^2}$	Wind-driven $\mathbf{E} \times \mathbf{B}$	$\dfrac{U}{L}$
∇p	$\dfrac{ne}{B}\left(\dfrac{kT}{e}\nabla n\right)$	Drift wave	$\dfrac{kT}{eL^2 B}$

The instability growth rate in each case is given by

$$\gamma = \frac{\text{velocity}}{L},$$

where L is the e-folding distance of the density perpendicular to
B. For instability, $\mathbf{J} \times \mathbf{B}$ must be parallel to ∇n. These processes create irregularities that can greatly affect communica-
tions and navigation signals traversing the ionosphere. The
most famous and debilitating effects are due to convective ionospheric storms (CISs; Kelley et al., 2011b).

6.5.1 Linear Theory of the Rayleigh-Taylor Instability

Dungey (1956) first proposed the Rayleigh-Taylor (R-T) instability as the process driving CISs. Historically, this process is called
equatorial spread F (ESF), which refers to the response of an ionosonde instrument to the irregularities. The output was then
"spread" in range and/or frequency at a given altitude. Here, we
prefer a physics-based nomenclature. This mechanism was tempo-
rarily rejected along with all of the other candidate theories by
Farley et al. (1970) since, as we shall see, it seemed only capable
of generating structure on the bottom side of the F region plasma
density profile. The manner in which the R-T instability can cause
irregularities to grow in the equatorial ionosphere is illustrated in
Fig. 6.3a using a two-dimensional model. Here, the steep, upward-
directed gradient that develops on the bottom side of the nighttime
F layer is approximated by a step function. The density is equal to
n_1 above the interface and zero below. The gravitational force is
downward, antiparallel to the density gradient, and the magnetic
field is horizontal, into the paper. An initial small sinusoidal perturbation is also illustrated, and we assume that the plasma is nearly
collisionless, that is, that κ_i and κ_e are large. We can determine the
electrical current by considering the ion and electron velocities
due to the pressure gradients and gravity. First, we note that the
pressure-driven current does not create any perturbation electric
field since the current is everywhere perpendicular to the density
gradient. The pressure-driven current thus flows parallel to the
modulated density pattern, has no divergence, and can be ignored.

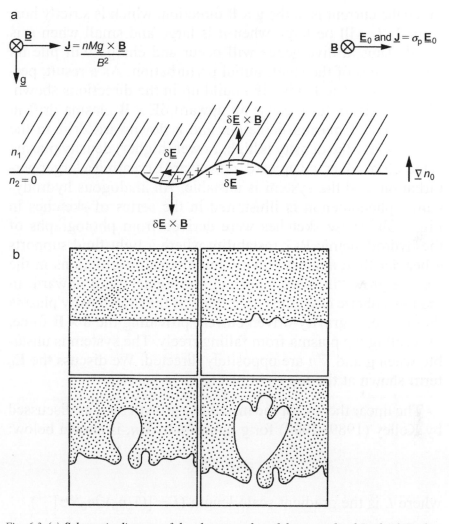

Fig. 6.3 (a) Schematic diagram of the plasma analog of the generalized Rayleigh-Taylor instability (including g and $E = E_0 + U \times B$) in the equatorial geometry. (b) Sequential sketches made from photos of the hydrodynamic Rayleigh-Taylor instability. A heavy fluid is initially supported by a transparent lighter fluid. After Kelley (2009). Reproduced with permission of Elsevier.

Turning to the gravitational term, the species velocity is proportional to its mass, and hence the ion term dominates. A net current flows in the x direction with magnitude

$$J_x = nMg/B.$$

Since the current is in the $\mathbf{g} \times \mathbf{B}$ direction, which is strictly horizontal, J_x will be large when n is large and small when n is small. Thus, a divergence will occur and charge will pile up on the edges of the small initial perturbation. As a result, perturbation electric fields ($\delta \mathbf{E}$) build up in the directions shown. These fields, in turn, cause an upward $\delta \mathbf{E} \times \mathbf{B}$ plasma drift in the region of plasma depletion and a downward drift in the region where the density is high. Lower (higher) density plasma is therefore advected upward (downward), creating a larger perturbation, and the system is unstable. An analogous hydrodynamic phenomenon is illustrated in the series of sketches in Fig. 6.3b. These sketches were derived from photographs of the hydrodynamic R-T instability where a light fluid supports a heavier fluid against gravity. Initial small oscillations in the surface grow "in place," pushing the lighter fluid upward. In the ionospheric case, the "light fluid" is the low-density plasma that carries a gravity-driven current providing the $\mathbf{J} \times \mathbf{B}$ force, preventing the plasma from falling freely. The system is unstable when $\mathbf{g}$ and ∇n are oppositely directed. We discuss the $\mathbf{E_0}$ term shown at the top right of Fig. 6.3a below.

The linear theory for the instability growth rate, as discussed by Kelley (1989, 2009) for g opposite to ∇n, is shown below:

$$\gamma = g/Lv_{\text{in}},$$

where L is the gradient scale length ($L = [(1/n_0)\partial n_0/\partial z]^{-1}$).

Although quite a simple expression, this result offers explanations for a number of CIS properties. First, in the initial development of spread F, there is a strong tendency for a VHF radar to obtain echoes confined to the height range where the density gradient is upward. In fact, the early Jicamarca study showed that the onset of nonthermal backscatter usually began at a density level of about 1% of the plasma density at the F peak. Several rockets have been flown during bottom side ESF at times when no radar echoes were obtained above the F peak and, indeed, intense irregularities were found below the peak, but a smooth profile was found above. These cases are thus in agreement with linear theory in that the latter predicts instability only when **g** is antiparallel to ∇n. Another feature predicted by the theory is a height dependency for γ due to the collision frequency term in the growth rate denominator: the higher the layer, the lower ν_{in} and the larger the instability growth rate. As mentioned earlier, Farley et al. (1970) noted a strong tendency for irregularities to be generated when the layer was at a high altitude where the collision term is small.

6.5.2 The Generalized R-T Process: Electric Fields, Neutral Winds, and Horizontal Gradients

Gravity is not the only destabilizing influence in the equatorial ionosphere. If we study the full current,

$$\mathbf{J}_\perp = \frac{neg}{\Omega_i} + \sigma_p(\mathbf{E} + \mathbf{U} \times \mathbf{B}),$$

we can include the effect of the ambient electric field $\mathbf{E}_0$ as well as the neutral wind. The latter can easily be included by remembering that when electric fields and winds both exist, the current density is $\sigma \cdot \mathbf{E}_0'$, where $\mathbf{E}_0' = \mathbf{E}_0 + \mathbf{U} \times \mathbf{B}$. Since the fundamental destabilizing source is the current, $\sigma_P \mathbf{E}'$ is the correct quantity to investigate. First, we note that, for a zonally eastward electric field, the zero-order Pederson current is in the same direction $(\mathbf{g} \times \mathbf{B})$ as the gravity-driven current. The derivation outlined above, which considers only gravity, can be generalized to include the effect of an electric field by replacing g/ν_{in} with $g/\nu_{in} + E_{x0}'/B$ in the growth rate where E_{x0}' is the zonal component of the electric field in the neutral frame of reference. A zonally eastward electric field drives a Pedersen current to the east. Any undulation of the boundary will intercept charge just as in the gravitational case and cause the perturbation to grow. Hence, an eastward E_{x0}' (eastward E_x and/or downward wind) is destabilizing. A zonally westward field will be stabilizing on the bottom side. The general condition for instability is that the $\mathbf{E}_{x0}' \times \mathbf{B}$ direction be parallel to the plasma density gradient. As discussed in the previous chapter, the zonal electric field component at the equator often increases to a large eastward value just after sunset, driving the F layer to very high altitudes. This uplift contributes in two ways to the destabilization of the plasma. Not only is the electric field in the right direction for instability but also the gravitational term becomes large due to the high altitude of the layer. The growth rates of the gravitational and electric field-driven processes are plotted as a function of height. For a 0.5-mV/m eastward E', the two sources of instability are equal at an altitude of 375 km. The gravitational term dominates above this height and increases exponentially with altitude (Fig. 6.4).

Since a large-scale neutral wind is usually horizontal, $(\mathbf{U} \times \mathbf{B}) \times \mathbf{B}$ is also horizontal and hence usually has no component parallel to ∇n if the ionosphere is vertically stratified. Hence, $(\mathbf{E}'_{x0} \times \mathbf{B}) \cdot \nabla n$ above is due entirely to the zonal electric field. However, other terms can be added to the linear growth rate by considering the possibility of a horizontal component of ∇n and/or a vertical wind. In fact, since the layer does

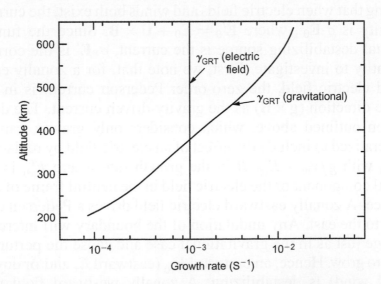

Fig. 6.4 Linear growth rates for the gravitational and electric field-driven Rayleigh-Taylor instabilities in the equatorial ionosphere for typical conditions. After Kelley et al. (1979). Reproduced with permission of the American Geophysical Union.

change height during the course of any given night, there is every reason to expect that the layer must be tilted with respect to the vertical. In such a case, the linear growth rate may be written in the following form:

$$\gamma_{RT} = \frac{E_x - wB}{LB} \cos \Delta + \frac{g}{v_{in}L} \cos \Delta + \frac{E_z + uB}{LB} \sin \Delta, \quad (6.6)$$

where L is the gradient scale length on the bottom side, Δ is the tilt angle, the z axis is upward, the x axis is eastward, and we ignore any dip angle of the magnetic field. If the tilt angle represents an inclined plane, a ball would roll down and west for positive Δ and down and east for negative Δ. This is the growth rate of the generalized R-T instability as applied to the equatorial ionosphere. Kelley et al. (1981), for example, used it to interpret Jicamarca data and concluded that the preference for plumes being generated during negative slopes of the Jicamarca radar profile was due to the contribution of an eastward neutral wind $(u\hat{\mathbf{a}}_x)$ to the instability growth rate. That is, if the ionosphere is tilted such that the neutral wind blows toward the east into a region of increasing plasma density so that $((u\hat{\mathbf{a}}_x \times \mathbf{B}) \times \mathbf{B})$ is antiparallel to the density gradient, a stable configuration occurs. However, if the wind blows antiparallel to the plasma gradient, which is believed to occur during the downward slopes, the wind is destabilizing. Tsunoda (1981, 1983), using the scanning radar at Kwajalein, also showed that wind effects arc important. Note, however, that if a perfect polarization field develops due to the F region dynamo, $E_z + uB = 0$, and no effect of the third term in Eq. (6.6) would occur. Remember that these instabilities are current driven, not electric field driven. However, due to a finite E-region loading of the current, u usually exceeds E_z/B by 20% or so, and a vertical current does flow. Since the electric field is generated by both local and remote wind fields (see Chapter 3), the generalized growth rate represents quite a complicated set of phenomena (we have not yet mentioned the role of shear in the plasma flow).

6.6 ELECTROJET INSTABILITIES

Two mechanisms can efficiently amplify thermal density fluctuations in the equatorial electrojet (EEJ). These mechanisms result from plasma instabilities known as the two-stream instability and the gradient drift (also known as the cross-field) instability, discussed in detail in Kelley (1989, 2009). The two-stream instability was discovered at Jicamarca in the early 1960s.

Many features of type 1 irregularities are explained by a modified two-stream instability theory developed independently by Farley (1963) using kinetic theory, and by Buneman (1963) using the Navier-Stokes equation. They show that the plasma is unstable for waves propagating in a cone of angle ϕ about the plasma drift velocity such that $V_D \cos \phi > C_s$. For smaller drift velocities, the plasma can still be unstable, provided there is a zero-order plasma density gradient oriented in the right direction relative to the electric field driving the electrons. This instability was first studied by Simon (1963) and Hoh (1963) for laboratory plasmas and is termed the gradient drift instability. Many features of type 2 radar echoes are explained by this instability.

Rocket data in a region of pure two-stream instability are shown in Fig. 6.5. The waves are seen at peaks of the sine wave due to $\mathbf{E} + \mathbf{V_R} \times \mathbf{B}$ where $\mathbf{E}$ is the field induced by rocket motion. The wiggles are the electrostatic field due to the two-stream process. The spectrum below shows that a sharp peak occurs at a wavelength of 2.5 m (the frequency plotted is due to the Doppler shift effect). Figure 6.5b shows the electric field structures in the height range where both the two-stream and gradient drift instabilities occur. The key feature is their square wave nature, which was also seen in radar data during the rocket flight (Kudeki et al., 1982).

The instability process now can be understood from Fig. 6.6, drawn for daytime conditions with the vertical electric field upward. The sinusoidal wave can thus represent both $\delta n / n$

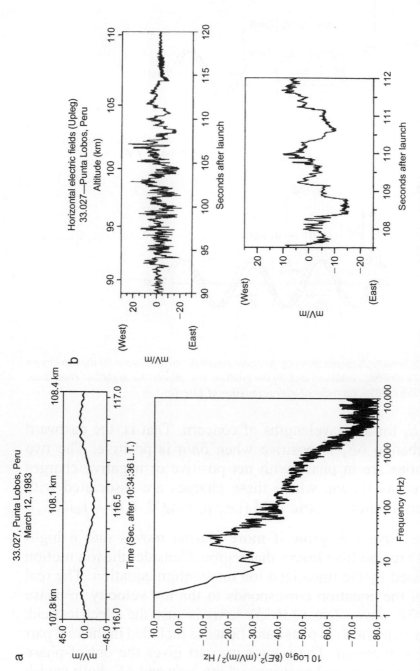

Fig. 6.5 (a) Raw data and the spectrum of pure two-stream waves in the daytime electrojet. The peak signal occurs when the detector is parallel to the current and the spectrum maximizes at a wavelength of a few meters. (b) Upleg electric field data with an expanded version of the horizontal electric fields. After Pfaff et al. (1987a,b). Reproduced with permission of the American Geophysical Union.

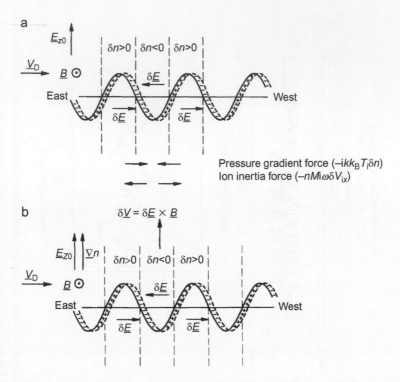

Fig. 6.6 Schematic diagrams showing the linear instability mechanism in (a) the two-stream process for daytime conditions and (b) the gradient drift process for nighttime conditions. After Kelley (2009). Reproduced with permission of Elsevier.

and δE_x for all wavelengths of concern. That is, the eastward perturbation δE_x is positive when $\delta n/n$ is positive. The two quantities are in phase with net positive or negative charges built up as shown, where these charges are associated with the perturbation electric field (i.e., $\rho_c = \varepsilon_0(\nabla \cdot \mathbf{E}) = -\varepsilon_0 ik\delta E_x$).

The wave will grow if more plasma moves into a high-density region than leaves that region. Consider the ion motion expressed by the linearized ion momentum equation. The real part of the equation corresponds to the ion velocity in phase with the wave. The $(+ik\delta\phi)$ term is just the electric field, δE_x, which is also in phase and hence is included in the real part of the expression. The imaginary part gives the out-of-phase motion at the spatial positions where $\delta n/n$ and δE_x both vanish and hence the part of the ion motion causing the wave to grow or decay. The two vectors, pressure gradient force and inertial force, are shown in Fig. 6.6a. For growth, the ion inertial force must be larger than the pressure term so that the plasma moves horizontally from low density to high density. This, in turn, requires that

$$\omega_r \delta V_{ix} > k(k_B T_i/M)(\delta n/n) \tag{6.7}$$

In other words, the ion inertial force must be greater than the pressure gradient force that is trying to smooth out the density enhancement by diffusion. Using Eq. (6.7) to eliminate δV_{ix}, the continuity equation,

$$i\omega(\delta n/n) - ik\partial V_{ix} = 0,$$

can also be written as

$$\omega_r \omega/k^2 > (k_B T_i/M).$$

Near marginal stability, $\gamma \ll \omega_r$, so $\omega \widetilde{\omega}_r$, and using the dispersion relation for ω_r/k, we have the requirement that (see Kelley, 2009 for details):

$$\frac{\delta E_x}{E_{z0}} = \frac{v_i}{\Omega_i(1 + \Psi_0)} \frac{\delta n}{n}, \tag{6.8}$$

$$V_D/(1 + \Psi_0) > (k_B T_i/M)^{1/2} \tag{6.9}$$

where $\Psi_0 = (v_{en} v_{in})/(\Omega_e \Omega_i)$. This is almost but not quite the required threshold condition for growth. The exact result, $V_D > (1 + \Psi_0)C_s$, came out of the detailed determinant analysis since the electron pressure term adds to the ion pressure via the ambipolar diffusion effect. Ion inertia is thus the destabilizing factor in the two-stream instability, whereas ambipolar diffusion causes damping.

Turning to the gradient drift instability, we must include the change of density due to advection from the vertical perturbation electron drift in the electron continuity equation. The term comes from $\mathbf{V} \cdot \nabla n$, which in linearized form adds a term $1/kL$ to the analysis where $L = [(1/n)(\mathrm{d}n/\mathrm{d}z)]^{-1}$ is the zero-order vertical gradient scale length. This added term changes only the imaginary part of ω; the real part is identical. This means that the relation of the perturbed field, δE_x, to the density perturbation, $\delta n/n$, for the gradient drift mode is the same for both instabilities. In studying the stability condition, consider Fig. 6.6b, which shows daytime conditions and includes a zero-order density gradient that is upward. From the figure, it is clear that the gradient is a destabilizing factor when it is upward since then the upward perturbation drift $\delta V_{ez} = \delta E_x/B$ occurs in a region where the density is already depleted ($\delta n < 0$). That is, a low-density region is convected upward into a region of higher background density, causing a growth in the relative value of $\delta n/n$. If the gradient is reversed in sign but E_{z0} remains the same, the perturbation electric fields will cause high-density plasma to drift into higher-density regions, which is stabilizing. Thus, for instability, the zero-order vertical electric field must have a component parallel to the zero-order density gradient. At the equator during normal daytime conditions, the unstable gradient direction is upward since E_{z0} is upward, whereas at night, the unstable gradient is downward.

This discussion has been a brief introduction to electrostatic waves in the ionosphere. More details may be found in Kelley (1989, 2009). Recently, many problems in nonlinear development were summarized by Kelley and Ilma (2013).

The EEJ has inspired thousands of papers, talks, and book chapters over the past 50 years that include the scientific fields of aeronomy, space plasma physics, pure plasma physics, and simulation/modeling challenges, as well as applications to space weather and geological prospecting (Fejer and Kelley, 1980; Kelley, 1989, 2009). Progress has been steady, but a few problems have been very difficult to solve. The most challenging questions are as follows:

1. Why is the phase velocity of plasma waves from EEJ type 1 radar echoes independent of the zenith angle, in

disagreement with the linear plasma theory for acoustic waves (first reported by Farley, 1963)?

2. Why do rocket-borne magnetic field measurements at EEJ heights disagree with theory (Gagnepain et al., 1977)?

3. Why are the vertical phase velocities of 3-m-wavelength plasma waves square wave in nature (Kudeki et al., 1982), as are the associated horizontal large-scale wave electric fields (Pfaff et al., 1987b)? See Fig. 6.5b for the electric field square waves.

Here, we review recent solutions to the first two questions and suggest possible solutions for the third.

6.6.1 A Quick Review of Questions 1 and 2

Kelley (2009) presented data that were relevant to all of these issues. Considering question 1 briefly, under normal equatorial conditions, radars see the same Doppler shift from zenith angles (θ) from 0° to 70°, which is equal to the sound speed, C_s. Linear theory of the pure Farley-Buneman instability predicts a $\sin(\theta)$ dependence (Farley, 1963). Kelley et al. (2008) showed that the sum of the vertical dc electric field, E_Z, and the wave-related electric field, δE, is such that $(\mathbf{E}_Z + \delta\mathbf{E}) \times \mathbf{B} / B^2$ is greater than C_s for all zenith angles in several radar range gates, since $\mathbf{E}_Z$ and $\delta\mathbf{E}$ are roughly equal in magnitude. As the wave phase velocity is equal or comparable to the electron line-of-sight drift (see the solution to question 3), the radar Doppler shift is independent of θ. Hysell et al. (2005) showed that the phase velocity is limited at the threshold for the two-stream instability and is equal to $C_s\cos\theta$. Thus, a large horizontal $\delta\mathbf{E}$ is needed at the equator.

Concerning question 2, classical EEJ theory predicts the peak height of the EEJ to be 4-5 km below the value measured on all seven existing rocket measurements. Previously, this was explained by arbitrarily increasing the electron-neutral collision frequency (ν_{en}) by a factor of four without explanation. Recently, Alken and Maus (2010b) appealed to the gradient drift instability to create an enhanced v_{en}, $v_{en}*$, as proposed by Ronchi et al. (1989). It is curious, though, that a nonlinear instability process would create $v_{en*} = 4v_e$ independent of the driving source of the instability: the zonal electric field. However, following the lead of Haerendel and Eccles (1992), Kelley et al. (2012) showed that, by using field-line-integrated conductivity, the peak height of the current could be explained classically without any nonlinear plasma physics. The model of Haerendel and Eccles (1992) did show an increase in the evening electrojet height, but they did not apply this identical result to the noontime electrojet controversy. Why? The field lines are curved, and even a 5-km height change makes the field line longer at 110 km than at 105 km. Note that the magnetic field data used by Alken and Maus (2010a) are from horizontal measurements on CHAMP and therefore could not predict the altitude of the peak current.

Question 3 was studied by Kelley and Ilma (2013), who suggested two possibilities. One is a wave-breaking process leading to nonlinear features. The second is generation of two-stream waves at the threshold for instability. The latter process would transfer the "growing" gradient drift wave energy to acoustic waves, which radiate that energy away.

6.7 ELECTROSTATIC WAVES AT OTHER LATITUDES

Entire chapters in Kelley (2009) are written about middle and high latitude instabilities. The equatorial theory is applicable but there are many differences due to the geometry and the dc electric field/wind patterns. The Perkins instability (Perkins, 1973) is new, as is the effect of turbulent mixing by solar wind/magnetospheric electric field fluctuations.

6.8 ON THE GENERAL APPLICATION OF FLUTE INSTABILITIES

Figure 6.7 dramatically shows how a plasma barium cloud is driven unstable by midlatitude neutral winds. The side of the cloud for which $U \times B$ is parallel to the plasma gradient is the unstable side. Figure 6.8 compares barium cloud structure to that of a high altitude (100 km) nuclear explosion. The features are essentially identical. The nuclear case is due to deceleration (g') of the expanding plasma by the background atmosphere, an R-T process. The two cases are identical if U (the wind velocity) is replaced by g'/v_{in} where v_{in} is the ion-neutral collision frequency.

The physics of the R-T instability is robust over 19 orders of spatial scale, as shown in Fig. 6.9 (Kelley et al., 2011a). The laboratory process is on the scale of 1 mm, the rocket experiment is 50 km, and the exploding star is 1 light year. The differences between the laboratory-developed and the other two are in the symmetry of their processes.

An example of the Perkins instability is presented in Fig. 6.10 (Otsuka et al., 2004). Here, simultaneous airglow images in two hemispheres are presented and are such that the dipole magnetic field lines connect the two ionospheres. The 630-nm emission is

Fig. 6.7 A photograph of a striated barium cloud taken at right angles to the magnetic field. After Kelley and Livingston (2003). Reproduced with permission of the American Geophysical Union.

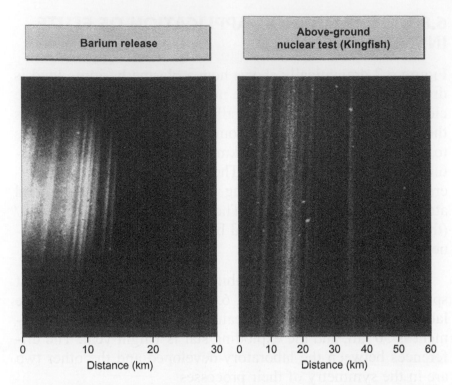

Fig. 6.8 *Comparison of a striated barium cloud and a late time nuclear explosion on the same distance scale. Both events occurred in the 150-200-km height range. After Kelley and Livingston (2003). Reproduced with permission of the American Geophysical Union.*

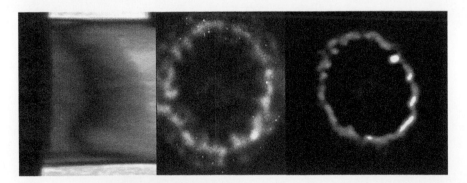

Fig. 6.9 *From left to right, a laboratory experiment, a rocket-borne space experiment, and a Hubble photo of the SN1987A supernova. (For a color version of this figure, the reader is referred to the online version of this chapter.) After Kelley et al. (2011a). Reproduced with permission of Scientific Research.*

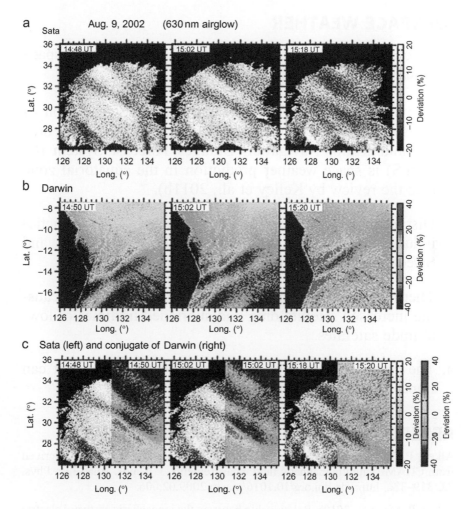

Fig. 6.10 Midlatitude structures on the same magnetic field lines. (For a color version of this figure, the reader is referred to the online version of this chapter.) After Otsuka et al. (2004). Reproduced with permission of the American Geophysical Union.

due to a combination of the ionospheric altitude and the plasma density. The former dominates and hence the dark regions are at high altitude and the bright at low altitudes. The altitude change is driven by a perturbation electric field. The angle of the features is thought to be due to the minimum damping value of gravity waves, which seed the instability (Kelley and Miller, 1997). The mirror images in both are due to mapping of electric fields between hemispheres.

6.9 SPACE WEATHER

Many of the processes discussed in this text contribute to space weather effects. Some examples are:

1. Communication/navigation outages due to scintillation of transionospheric radio waves. The goal of the Air Force Communication/Navigation Outages Forecast System (C/NOFS) is space weather prediction in the equatorial zone (see the review by Kelley et al., 2011b).

2. Disruption of electrical power systems by geomagnetic-induced current. Quebec lost power for 10 h in 1989, and Sweden and South Africa have had similar outages.

3. Thermospheric Joule heating expands the atmosphere, causing increased satellite drag and shortened lifetimes of low-altitude satellites.

4. The creation of intense energetic electron fluxes that can impact satellite lifetime.

REFERENCES

Alken, P., Maus, S., 2010a. Electric fields in the equatorial ionosphere derived from CHAMP satellite magnetic field measurements. J. Atmos. Solar-Terr. Phys. 72, 319–326. http://dx.doi.org/10.1016/j.jastp.2009.02.006.

Alken, P., Maus, S., 2010b. Relationship between the ionospheric eastward electric field and the equatorial electrojet. Geophys. Res. Lett. 37, L04104. http://dx.doi.org/10.1029/2009GL041989.

Buneman, O., 1963. Excitation of field aligned sound waves by electron streams. Phys. Rev. Lett. 10 (7), 285–287.

Dungey, J.W., 1956. The effect of ambipolar diffusion in the night-time F layer. J. Atmos. Terr. Phys. 9, 90–102.

Farley, D.T., 1963. A plasma instability resulting in field-aligned irregularities in the ionosphere. J. Geophys. Res. 68 (22), 6083–6097. http://dx.doi.org/10.1029/JZ068i022p06083.

Farley, D.T., Balsley, B.B., Woodman, R.F., McClure, J.P., 1970. Equatorial spread F: implications of VHF radar observations. J. Geophys. Res. 75 (34), 7199–7216. http://dx.doi.org/10.1029/JA075i034p07199.

Fejer, B.G., Kelley, M.C., 1980. Ionospheric irregularities. Rev. Geophys. 18 (2), 401–454.

Gagnepain, J., Crochet, M., Richmond, A.D., 1977. Comparison of equatorial electrojet models. J. Atmos. Solar-Terr. Phys. 39, 1119–1124.

Haerendel, G., Eccles, J.V., 1992. The role of the equatorial electrojet in the evening ionosphere. J. Geophys. Res. 97, 1181–1192. http://dx.doi.org/10.1029/91JA02227.

Hoh, F.D., 1963. Instability of Penning-type discharges. Phys. Fluids. 6 (8), 1184–1191. http://dx.doi.org/10.1063/1.1706878.

Hysell, D.L., Larsen, M.F., Swenson, C.M., Barjatya, A., Wheeler, T.F., Sarango, M.F., Woodman, R.F., Chau, J.L., 2005. Onset conditions for equatorial spread F determined during EQUIS II. Geophys. Res. Lett. 32 (24), L24104. http://dx.doi.org/10.1029/2005GL024743.

Kelley, M.C., 1989. The Earth's Ionosphere: Plasma Physics and Electrodynamics, International Geophysics Series, vol. 43. Academic Press, San Diego, CA.

Kelley, M.C., 2009. The Earth's Ionosphere: Plasma Physics and Electrodynamics, second ed. International Geophysics Series, vol. 96. Academic Press, Burlington, MA.

Kelley, M.C., Ilma, R.R., 2013. On the development of nonlinear waves in the equatorial electrojet. J. Atmos. Solar-Terr. Phys. 103, 3–7. http://dx.doi.org/10.1016/j.jastp.2012.12.009.

Kelley, M.C., Livingston, R., 2003. Barium cloud striations revisited. J. Geophys. Res. 108 (A1), 1044. http://dx.doi.org/10.1029/2002JA009412.

Kelley, M.C., Miller, C.A., 1997. Electrodynamics of midlatitude spread F, 3, Electrohydrodynamic waves? A new look at the role of electric fields in thermospheric wave dynamics. J. Geophys. Res. 102 (A6), 11539–11547.

Kelley, M.C., Baker, K., Ulwick, J., 1979. Late time barium cloud striations and their possible relationship to equatorial spread F. J. Geophys. Res. 84, 1898–1904.

Kelley, M.C., Larsen, M.F., LaHoz, C., McClure, J.P., 1981. Gravity wave initiation of equatorial spread F: a case study. J. Geophys. Res. 86 (A11), 9087–9100. http://dx.doi.org/10.1029/JA086iA11p09087.

Kelley, M.C., Baker, S.D., Holzworth, R.H., Argo, P., Cummer, S.A., 1997. LF and MF observations of the lightning electromagnetic pulse at ionospheric altitudes. Geophys. Res. Lett. 24 (9), 1111–1114. http://dx.doi.org/10.1029/97GL00991.

Kelley, M.C., Cuevas, R.A., Hysell, D.L., 2008. Radar scatter from equatorial electrojet waves: an explanation for the constancy of the Type I Doppler shift with zenith angle. Geophys. Res. Lett. 350, L04106. http://dx.doi.org/10.1029/2007GL032848.

Kelley, M.C., Dao, E., Kuranz, C., Stenbaek-Nielsen, H., 2011a. Similarity of Rayleigh-Taylor Instability development on scales from 1 mm to one light year. Int. J. Astron. Astrophys. 1 (4), 173–176.

Kelley, M.C., Makela, J., de La Beaujardiere, O., Retterer, J., 2011b. Convective ionospheric storms: a review. Rev. Geophys. 49 (2), RG2003. http://dx.doi.org/10.1029/2010RG000340.

Kelley, M.C., Ilma, R.R., Eccles, V., 2012. Reconciliation of rocket-based magnetic field measurements in the equatorial electrojet with classical collision theory. J. Geophys. Res. 117 (A1), A01311. http://dx.doi.org/10.1029/2011JA017020.

Kudeki, E., Farley, D.T., Fejer, B.G., 1982. Long wavelength irregularities in the equatorial electrojet. Geophys. Res. Lett. 9 (6), 684–687. http://dx.doi.org/10.1029/GL009i006p00684.

Otsuka, Y., Shiokawa, K., Ogawa, T., Wilkinson, P., 2004. Geomagnetic conjugate observations of medium-scale traveling ionospheric disturbances at midlatitude using all-sky airglow imagers. Geophys. Res. Lett. 31, L15803. http://dx.doi.org/10.1029/2004GL020262.

Perkins, F.W., 1973. Spread F and ionospheric currents. J. Geophys. Res. 78 (1), 218–226.

Pfaff, R.F., Kelley, M.C., Kudeki, E., Fejer, B.G., Baker, K.D., 1987a. Electric field and plasma density measurements in the strongly driven daytime equatorial electrojet: 1. The unstable layer and gradient drift waves. J. Geophys. Res. 92 (A12), 13578–13596. http://dx.doi.org/10.1029/JA092iA12p13578.

Pfaff, R.F., Kelley, M.C., Kudeki, E., Fejer, B.G., Baker, K.D., 1987b. Electric field and plasma density measurements in the strongly-driven daytime equatorial electrojet: 2. Two-stream waves. J. Geophys. Res. 92 (A12), 13597–13612.

Ronchi, C., Similon, P.L., Sudan, R.N., 1989. A nonlocal linear theory of the gradient drift instability in the equatorial electrojet. J. Geophys. Res. 94 (A2), 1317–1326. http://dx.doi.org/10.1029/JA094iA02p01317.

Simon, A., 1963. Instability of a partially ionized plasma in crossed electric and magnetic fields. Phys. Fluids. 6 (3), 382–388. http://dx.doi.org/10.1063/1.1706743.

Stix, T.H., 1962. The Theory of Plasma Waves. McGraw-Hill, New York.

Stix, T.H., 1992. Waves in Plasmas. Springer-Verlag, New York.

Tsunoda, R.T., 1981. Time evolution and dynamics of equatorial backscatter plumes, 1. Growth phase. J. Geophys. Res. 86 (A1), 139–149. http://dx.doi.org/10.1029/JA086iA01p00139.

Tsunoda, R.T., 1983. On the generation and growth of equatorial backscatter plumes, 2. Structuring of the west walls of upwellings. J. Geophys. Res. 88 (A6), 4869–4874. http://dx.doi.org/10.1029/JA088iA06p04869.

Electric Field Measurement Techniques

The Earth's Electric Field. http://dx.doi.org/10.1016/B978-0-12-397886-8.00007-7

7.1 INTRODUCTION

Measurements of atmospheric electric phenomena have been conducted for over a hundred years, going back to the experiments of Benjamin Franklin. These experiments were carried out in a very weakly ionized medium. Advances during the past few decades have been carried out in a highly conducting material called a plasma. New methods were needed and developed in the 1960s, with the realization of electric phenomena's importance in space. As we shall see, the field can be deduced by direct measurements using what is basically a voltmeter or by measuring the plasma motion in response to electrical forcing. The latter has been accomplished by using chemical tracers, whose motion can be registered by cameras, by radars on the ground that record the material's Doppler shift, and by detecting the plasma motion relative to a satellite's velocity.

7.2 BASIS FOR THE MEASUREMENT TECHNIQUES

The relationship between the electric field and the plasma velocity allows different measurement techniques to be applied in determining the field. This, in turn, is related to the equation,

$$\mathbf{V}_{\text{perp}} = \frac{\mathbf{E} \times \mathbf{B}}{B^2}$$

where, due to the cross product, the velocity in this expression is perpendicular to the magnetic field. Except for the source of the particle acceleration, which causes the aurora, the parallel component of the electric field is very small and essentially not measureable. This is not true of the plasma velocity, however, which can have a large magnitude along the magnetic field that is comparable to the perpendicular velocity. Thus, measuring the velocity provides more information on the medium than directly measuring the electric field does. The measurement techniques we describe below thus complement each other.

We deal with the four most important measurement methods below:

1. Incoherent scatter radar (ISR) measurements using large aperture, high-power systems from ground level.

2. Double-probe measurements on rockets and satellites.

3. Plasma drift measurements on satellites.

4. Camera triangulation of ion cloud releases (barium vapor) from rockets and satellites.

The fact that direct electric field measurements are possible on rockets, unlike plasma drift measurements, has important implications. First, these measurements facilitate lower thermospheric studies that are not possible with satellite sensors. As many important phenomena occur in this region, rocket use has been crucial in understanding the lower thermosphere. Second, launching rockets with countdown times as short as 3 min and, in rare cases, even less, allows phenomena that are localized in time and space to be studied in great detail. An example is the study of auroral processes. Dozens of rockets have been launched successfully over auroral displays, which are highly dynamic. Another example is the use of ground-based radars to first identify processes of interest and then launch into and over these regions of interest. Rockets have been flown in conjunction with radar data from eight such systems over the years. Also, the data rates from rocket data of all types taken by these short-lived spacecraft are usually two orders of magnitude higher than would be possible using satellites. As a number of radar sites are not collocated with permanent rocket launch sites, NASA has carried out several campaigns at sites in Peru, Greenland, and Puerto Rico.

7.3 ISR TECHNIQUES

The ground was broken for an ISR site in Arecibo, Puerto Rico, in 1959 and nearly simultaneously at Jicamarca, Peru. At the time of writing, there are eight such sites in regular operation outside the former Soviet Union and two inside (Ukraine and Siberia).

Like an ionosonde, an ISR transmits a radio wave signal and receives a returned "echo" sometime later. Ionosondes operate over frequencies typically in the range of 1-20 MHz and rely on the fact that such a signal is reflected when its frequency f is equal to the local plasma frequency f_p where

$$f_p = (2\pi)^{-1} \left(ne^2 / m_g \varepsilon_0 \right)^{1/2},$$

which, to a good approximation, is

$$f_p = 9000 \sqrt{n} \, \text{Hz}$$

where n is the plasma density in reciprocal cubic centimeters. As the peak plasma density in the ionosphere is a few times $10^6 \, \text{cm}^{-3}$, $f_p \leq 12 \, \text{MHz}$. Vertically transmitted frequencies above the peak plasma frequency in the overhead ionosphere go right through. Consequently, ground-based ionosondes yield no information above the height of the peak on the F-region plasma density. If there is a dense E layer, it can block the F layer entirely. Occasionally, due to metallic ions, a thin low-altitude layer will allow some energy to tunnel through, revealing the F layer above.

The ISR technique circumvents both the reflection and transmission problems by using (a) frequencies well above any reasonable plasma frequency in the natural ionosphere and (b) very sensitive systems. The lowest frequency used for ISR is about 50 MHz. Such waves are virtually unattenuated by the ionosphere and pass through almost unaffected into space. "Almost" is the key word here, as it is the small amount of energy scattered by the ionospheric electrons that is used by the ISR method. To gain some perspective, the amount of energy scattered back to a typical ISR antenna from 300 km is roughly comparable to the target represented by a small copper coin at the same range. It should be no surprise, then, that the largest of the ISR sites measures their effective power-aperture product in tens of megawatt-acres or, in more reasonable units, a few times 10^{11} W m^2. The two most powerful systems—at Arecibo, Puerto Rico and Jicamarca, Peru—have antennas roughly 1 km in circumference and transmit megawatts of power.

As with most radar systems, the signal is transmitted in pulses with range from the radar to the echoing region determined by dividing half the delay time by the speed of light. As the ionosphere is an extended target, various ranges can be interrogated using the same set of pulses by analyzing the returned signal at appropriate time delays.

The information in the returned signal is remarkably rich in physical content. The power in the returned echo is proportional to the number density of electrons in the volume irradiated. This stems from the fact that each electron incoherently radiates back a small amount of the incident energy. The electric field in the transmitted wave causes any and all electrons encountered during transit of the pulse to oscillate, radiating a signal at almost the same frequency. This fact alone was sufficient to generate the enthusiasm necessary to build the first site at Arecibo, as, for the first time, information on the ionosphere would be available above the F peak. This ability was developed around the same time as when satellites were exploring the same region. The richness in information content stems as much from the spectrum of the returned signal as from the power returned, however. The electrons are consistently in thermal motion and hence the radiation is Doppler-shifted from the incident frequency. The result is a spread in the returned radio wave spectrum that contains considerable information about the velocities present in the medium and hence about its dynamics. Near-simultaneous development of ISR and satellite probes was crucial, as the theoretical basis for the former was very strong and helped calibrate the *in situ* instruments, which had a less solid theoretical basis.

The original idea (Gordon, 1958) hypothesized quite naturally that the Doppler width of the returned spectrum would be of the order

$$\delta f / f \cong \langle V \rangle_e / c$$

where $\langle V \rangle_e$ is the mean thermal electron velocity. For typical parameters, δf would be a megahertz. Such a bandwidth permits a huge amount of cosmic noise to mask the signal and requires very large systems. Surprisingly, the first experiments by Bowles (1958) showed the Doppler width to be of the order of $\langle V \rangle_i / c$, where $\langle V \rangle_i$ is the ion thermal speed. This allowed much smaller systems to be deployed after the 1960s. This result is explained by the fact that, when the probing wavelength is greater than the Debye length, the Debye cloud around each electron and the ions inside that cloud all contribute to the scattering, which slows down the scattering process. The Debye length is a characteristic distance over which ions and electrons can be separated in a plasma (Chen et al., 1984) and is equal to the ratio of the electron thermal velocity divided by the plasma frequency. In the ionosphere, typical values range from 1 cm to 1 m.

A schematic diagram of the Doppler spread due to scattering from thermal fluctuations in the so-called ion line is presented in Fig. 7.1. The power under the curve is proportional to the number density, and the width δf can be used to determine

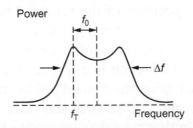

Fig. 7.1 Schematic diagram of the Doppler spectrum associated with backscatter from thermal fluctuation in the upper atmosphere. Here f_T is the transmitted frequency, f_0 is the mean Doppler shift, which yields the line-of-sight velocity, and Δf is a measure of the width of the spectrum. Only the ion line is shown here. After Kelley (2009). Reproduced with permission of Elsevier.

the ion temperature. The relative intensity of the "wings" in the spectrum yields the electron temperature, and the overall shift of the spectrum, f_0, from the transmitted frequency f_T gives the line-of-sight component of the mean ion velocity $\mathbf{V}_i$. If several positions are used, the complete vector flow velocity can be found. In the F region, the ion flow velocity yields the perpendicular electric field components via the relationship

$$\mathbf{E}_\perp = \mathbf{B} \times \mathbf{V}_i.$$

The ion flow velocity parallel to $\mathbf{B}$ is much more complicated as many factors contribute, such as the component of the neutral wind along $\mathbf{B}$, the ion pressure gradient, and gravitational forces. However, this information is important for understanding thermospheric winds, and much effort has gone into analyzing radar data to separate these various sources of parallel flow (Kelley, 2009).

The theory of scattering from thermal fluctuations is very well understood. In fact, as pointed out by Farley (1979), when viewed as a test of the crucial Landau damping method for carrying out Vlasov plasma theory, measurements of the incoherent scatter spectrum have verified that theory to better than 1% accuracy. The radar scatters from irregularities in the medium, $\mathbf{k}_m$, according to the relationship

$$\mathbf{k}_T = \mathbf{k}_s \times \mathbf{k}_m \tag{7.1}$$

where $\mathbf{k}_T$ is the transmitted wave and $\mathbf{k}_s$ is the scattered wave. As $\mathbf{k}_s = -\mathbf{k}_T$ for backscatter,

$$\mathbf{k}_m = 2\mathbf{k}_T$$

or

$$\lambda_m = \lambda_T/2.$$

Thus, the scattering wavelength is one-half the transmitted wavelength. Equation (7.1) represents conservation of momentum as each photon carries a momentum equal to $\hbar \mathbf{k}$ where $\hbar$ is Planck's constant. Conservation of energy (as each photon energy equals $\hbar \omega$) requires

$$\omega_T = \omega_s + \omega_m. \tag{7.2}$$

Thus, the Doppler width of the returned spectrum ($\omega_T - \omega_s$) is related to the frequencies of waves in the medium, ω_m. As thermal fluctuations can be considered as a superposition of damped sound waves, the spread in ω_m is of the order

$$\omega_m \approx |\mathbf{k}_m| C_s$$

where C_s is the sound speed in a plasma,

$$C_s^2 = (kT_e + kT_i)/M_i.$$

The width of the ion line for a 50-MHz radar in a plasma with $C_s \approx 1000$ m/s is thus of the order $[(4\pi/6)\,\mathrm{m}]$ $(1000\,\mathrm{m/s}) \approx 4000\,\mathrm{rad/s} \approx 600$ Hz. For a medium of the same temperature, the Doppler width will just scale with the radar frequency, so at Arecibo (430 MHz), the width of the "ion line" is about eight times larger. These frequencies are the ratio m_e/M_i smaller than would occur for radar wavelengths smaller than the Debye length. This explains the smaller bandwidth of the scattered signal.

The peaks or wings in the ion line spectrum correspond to the frequency of the normal modes in the medium, that is, the frequency at which a finite-amplitude sinusoidal sound wave of wave number $\pm \mathbf{k}_m$ would exist. In thermal equilibrium, such waves are damped. The time dependence associated with this damping fills in the remainder of the spectrum above and below $|\omega_s| = |\mathbf{k}_m| C_s$. When $T_e/T_i > 1$, Vlasov theory predicts that the sound wave damping will be less. These wings increase in amplitude relative to the rest of the spectrum when T_e/T_i increases.

In ion sound waves, both electrons and ions participate. The electron gas alone can itself support thermal fluctuations, which, at the scattering wave number, occur at frequencies of the order $\mathbf{k}_m\langle V\rangle_e$. This frequency is larger by a factor of M_i/m_e than the frequency of the wings in the ion line. In the ionosphere for O^+, this ratio is about 3×10^4, so the pure electron waves are Doppler-shifted considerably more than the mixed ion-electron waves. The first simultaneous observations of both ion and electron scattering lines are reproduced in Fig. 7.2. The data were taken in the altitude range of 1460-2486 km over Arecibo, where H^+ is the dominant ion. The electron line has nearly the same power as the ion line but is spread over a much wider bandwidth. This makes the signal-to-noise ratio much smaller, and far fewer measurements of this scattering source exist. The narrow nature of the ion component is extremely crucial to detecting the small Doppler shift caused by plasma motion.

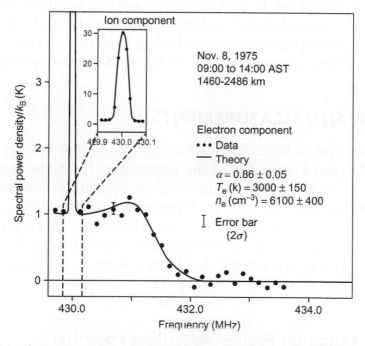

Fig. 7.2 Incoherent scatter data from the ionosphere at Arecibo showing, for the first time, the electronic as well as ionic portions of the spectrum. Due to the wide bandwidth and resulting low signal-to-noise ratio, a great deal of averaging in both time and space is required. After Hagen and Behnke (1976). Reproduced with permission of the American Geophysical Union.

In practice, most ISR systems measure the temporal autocorrelation function (ACF) of the signals scattered from a fixed point in space by sending up a series of pulses. As the frequency spectrum is the Fourier transform of the ACF, the spectrum can be generated at will on the ground using the ACF. As computational power has increased over the years, advancements in computational power have dominated developments in the field. The temporal resolution is then determined by how many such measurements must be averaged to generate a reasonable ACF. As the ISR theory is exact, least-squares fitting methods can be used on the ACF waveform, including second-order ion species and differential drifts of the various ions. This further increases the ionospheric information available via the ISR method. Examples of ACF measurements are presented in Fig. 7.3 as a function of altitude over Arecibo. The fitted theoretical ACFs are superposed on the measurements and can hardly be distinguished. Notice that, as the altitude increases, the ACF narrows in time, corresponding to the widening of the Doppler spectrum as the plasma changes from O^+ to H^+ with increasing altitude.

7.4 *IN SITU* MEASUREMENTS

Satellite and rocket-borne instruments of all kinds have greatly added to our knowledge of the ionosphere. Here, we concentrate on techniques that are used to determine ionospheric parameters such as density, temperature, drift velocity, and electric fields. The treatment is not exhaustive but is representative of some of the instrumentation referred to in the text. Other reviews of interest have been written by Bauer and Nagy (1975) and Mozer (1973).

7.4.1 Langmuir Probes, Retarding Potential Analyses, and Drift Meters

In this section, we deal with a variety of instruments used to measure the temperature, concentration, and drift velocity of

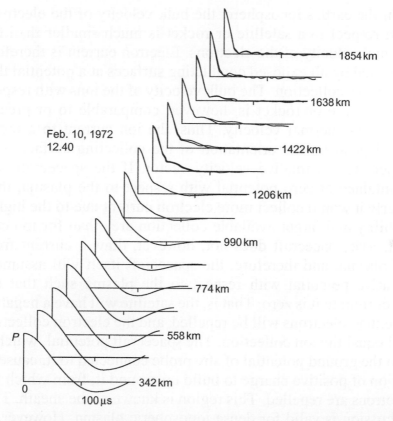

Feb. 10, 1972
12.40

1854 km

1638 km

1422 km

1206 km

990 km

774 km

558 km

342 km

P

0

100 μs

Fig. 7.3 Incoherent scatter autocorrelation function measurements made at Arecibo. The pulse length used was 2 ms and the integration time was 20 min. The experimental and fitted theoretical curves are plotted together, but the agreement is so good that the two often cannot be distinguished. After Hagen and Hsu (1974). Reproduced with permission of the American Geophysical Union.

either the ambient thermal electrons or the thermal ions. These instruments are mounted on satellites and rockets that are moving through the plasma at velocities between 1 and 9 km/s. In such cases, any conducting surface will collect an electron current and an ion current that, in most cases, can be calculated by assuming that the plasma has a drifting Maxwellian distribution function. If the conductor is held at some potential P, the current is calculated by integrating the distribution function over the surface area of the collector for all energies greater than P.

In the earth's ionosphere, the bulk velocity of the electrons with respect to a satellite or rocket is much smaller than the thermal velocity of the electrons. Electron current is therefore collected by all exposed conducting surfaces at a potential that will allow collection. The bulk velocity of the ions with respect to a satellite or rocket is, however, comparable to or greater than their thermal velocity. Thus, the ion current to a probe can depend on the orientation of the collecting surface with respect to the relative velocity vector. If the spacecraft was maintained at zero potential with respect to the plasma, then clearly it would collect more electron current due to the higher mobility and larger available collection area than for ion current. The spacecraft does not, however, draw a current from the plasma, and therefore, the spacecraft itself will assume a negative potential with respect to the plasma such that the net current to it is zero. That is, the satellite will have a negative potential, electrons will be repelled, and the electron collection will equal the ion collection. The spacecraft potential, which is also the ground potential of any probe connected to it, causes a region of positive charge to build up around it, from which the electrons are repelled. This region is known as the sheath. This discussion is valid for dense ionospheric plasma. However, in deep space, photoemission from the satellite can exceed the electron flux and the spacecraft will charge positively. A second case occurs in very high temperature/low density regions (e.g., the Van Allen belts) during an eclipse when spacecraft can reach negative voltages of kilowatts and cause severe damage.

7.4.2 The Langmuir Probe

Mott-Smith and Langmuir (1926) published a classical paper on the current collection properties of a probe in a plasma from which the modern Langmuir probe derives its name. Most Langmuir probes consist of small conducting surfaces with cylindrical or spherical geometries. Typical sensor dimensions are a 2-cm-diameter sphere or a 20-cm-long, 0.2-cm-diameter cylinder. The probes are usually mounted on short booms of about 20-cm length to project beyond the spacecraft sheath. As mentioned before, the current collection properties of these probes depend on the shape and area of the collector. As we shall see, however, they do not have a large effect on the determination of electron temperature. The instruments function by applying a varying voltage to the probe that covers the range of energies of interest. Typical probe voltages may vary between +5 and −5 V. Figure 7.4 shows a characteristic curve of probe current versus voltage that one might obtain from a cylindrical Langmuir probe. In fact, all Langmuir characteristic curves have the same region. The ion saturation region occurs where the probe potential is sufficiently negative to repel all the thermal electrons and begin to attract the ions. The electron retardation region exists where the ion current is not affected greatly by the potential on the probe, but some electrons are repelled. Finally, the electron saturation region exists where all the

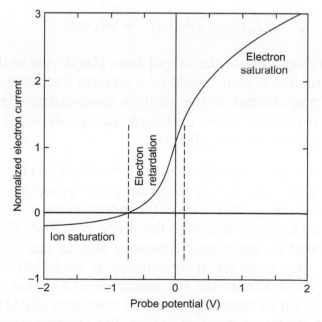

Fig. 7.4 A typical Langmuir probe I-V characteristic. After Kelley (2009). Reproduced with permission of Elsevier.

thermal electrons are attracted to the probe and where the probe potential repels the ion current. In the electron retardation region, the probe current is given by

$$I_e = N_e q A (k_B T_e / 2\pi m_e)^{1/2} \exp(-q\Phi / k_B T_e)$$

where A is the collector area, N_e is the electron concentration, q is the electron charge, k_B is the Boltzmann constant, T_e is the electron temperature, m_e is the electron mass, and Φ is the retarding potential applied to the probe.

A variety of techniques can be employed to extract the electron temperature from such a characteristic curve. If a least-squares fit to the original data can be performed, then either knowledge of the electron concentration is required or this parameter must also be a least-squares variable. An alternative lies in considering the logarithm of the electron current rather than the current itself. Taking the logarithm of both sides of the previous equation,

$$\ln(I_e) = -q\Phi / k_B T_e + \ln(\text{const.}).$$

Thus, the logarithm of the output from this device in the electron retardation region should be a straight line for which the slope is proportional to the electron temperature. Using this result, it is possible to use on-board microprocessors to derive the electron temperature directly.

In the two saturation regions, the ion current and the electron current can be obtained by integrating the appropriate drifting Maxwellian energy distribution function over the collector surface for all energies less than the probe potential. These currents depend on the sensor geometry and, as can be seen in Fig. 7.4, they need not truly saturate at potentials less than 5 V. We will deal with the ion saturation of a planar probe in the discussion of retarding potential analyzers (RPAs). In the so-called electron saturation region, the electron current to a cylindrical collector is given approximately by

$$I_e = 2AN_e q \left(\frac{k_B T_e}{2\pi m_e}\right)^{1/2} \frac{1}{\pi^{1/2}} \left(1 + \frac{q\Phi}{k_B T_e}\right)^{1/2}.$$

This expression is derived from the assumption of an infinitely long cylinder and therefore has an "end effect" correction that is usually small for typical cylindrical probe dimensions (Szuszczewicz and Takacs, 1979). Having derived the electron temperature from the electron retardation region, it can be seen that the electron saturation portion of the curve can be simply used to derive the electron concentration. In fact, the ion saturation portion of the curve can be similarly used to calculate the total ion concentration. The latter has the advantage that when the probe velocity is much greater than the thermal velocity of the ions, the expression for the total ion concentration is essentially independent of the ion temperature and a very small end effect correction. As the ionospheric plasma is charge neutral, this approach is frequently employed to derive the total plasma concentration from a Langmuir probe.

7.4.3 Ion Temperature, Density, and Drift Velocity Measurements: The RPA

RPAs measure the ambient ion current to a collector as a function of an applied retarding potential. In a manner similar to a Langmuir probe, a curve of ion current versus retarding potential is obtained from which the thermal ion temperature can be determined.

The velocity of an earth-orbiting satellite can be as high as 8 km/s in the ionosphere. At this velocity, the kinetic energy of the ions ($1/2\ mV^2$) is equivalent to 1/3 eV/amu. The presence of molecular and metallic ions (Fe^+) in the ionosphere therefore requires that retarding potentials between 20 and 30 V be applied in the sensor. An RPA sensor typically has a planar or spherical geometry. A spherical geometry has the advantage that it can be mounted on a spacecraft in such a way that it has access to the flowing ion gas for most orientations of the spacecraft. The plasma sensor must, however, be mounted to view along the velocity vector of the satellite, thus precluding measurements more than once per spin period if the vehicle is spinning. The planar sensor has the advantage that its known orientation with respect to the spacecraft velocity allows additional information to be extracted from the curve of ion current versus retarding potential. In addition, the construction of a planar geometry leads to more uniform electric potentials in the sensor. From the standpoint of *velocity* measurements, the RPA can be used to determine its component parallel to the satellite velocity vector, albeit velocity with a low time/space resolution due to the sweep needed.

7.4.4 Ion Drift Velocity Measurements: The Ion Drift Meter

We have seen that a planar RPA is capable of measuring the component of the ion drift velocity along the axis of the sensor. In the earth's ionosphere, where an orbiting spacecraft moves at a velocity much larger than the ion thermal speed, a rather simple device called an ion drift meter (IDM) can be used to measure the other two mutually perpendicular ion drift velocity components. The IDM has a planar geometry similar to the RPA but has a square entrance aperture and a segmented collector, as shown schematically in Fig. 7.5.

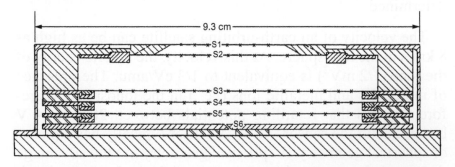

Fig. 7.5 Modification of a plasma ion sensor to measure the plasma drift velocity. After Kelley (2009). Reproduced with permission of Elsevier.

The grid S5 is biased negatively to prevent ambient electrons from reaching the collector and to suppress any photoelectrons liberated from the collector surface. All other grids are grounded to ensure a field-free region between S2 and S3 through which the ambient ions can drift. As the ions are moving supersonically with respect to the sensor, they form a collimated beam in the manner shown in Fig. 7.6. The collector area that they illuminate will therefore depend on the angle at which they arrive at the sensor. As the current collected on each collector segment is proportional to the area struck by the ion beam, it can be shown quite easily that the ratio of the currents of two segments is proportional to the tangent

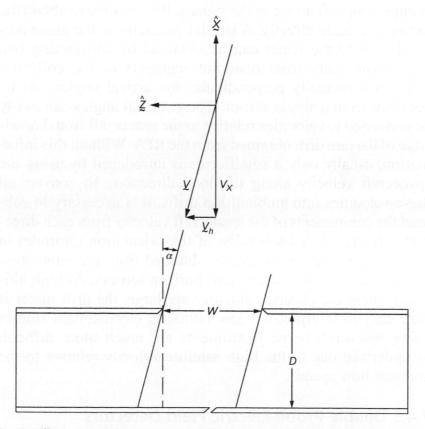

Fig. 7.6 Illustration of the geometry involved in the interpretation of measured particle arrival angles. After Kelley (2009). Reproduced with permission of Elsevier.

of the ion arrival angle. If the entrance aperture has straight edges and the currents to each collector pair are denoted by I_1 and I_2, then

$$\log I_1 - \log I_2 = \log(I_1/I_2)$$
$$= \log[(H + D\tan\alpha)/(H - D\tan\alpha)].$$

If we let $I_1 - I_2 = \Delta I$ and $I = I_1 + I_2$, then to first order, in $\Delta I / I$, we can show that

$$\log I_1 - \log I_2 = (2H/D)\tan\alpha.$$

Thus, by using a combination of logarithmic amplifiers to measure the collector current and a linear difference amplifier to supply the difference in the signals, it is possible to obtain the ion arrival angle directly. A similar geometry in the plane perpendicular to the paper can be obtained by considering two appropriate pairs from quadrant segments of the collector. Thus, two mutually perpendicular ion arrival angles can be obtained from a single sensor. These arrival angles can easily be converted to velocities relative to the spacecraft from knowledge of the ram drift obtained from the RPA. Without this information, usually only a small error is introduced by using the spacecraft velocity along the look direction. To convert all these velocities into ambient ion drifts, it is necessary to subtract the components of the spacecraft velocity from each direction. This requires knowledge of the orientation (attitude) in inertial space, which is usually obtained from a combination of star sensors, sun sensors, and horizon sensors. At high altitudes where the plasma velocities are large, the drift meter is very easy to interpret. At low latitudes, electric field studies using spacecraft-borne instruments are much more difficult to undertake due to the high satellite velocity relative to the ambient flow speeds.

7.4.5 Double-Probe Electric Field Detectors

The double-probe technique has been used successfully to measure electric fields on balloons, rockets, and satellites. In essence, at dc, the technique makes a resistive contact to the plasma at two separated positions. If the two electrodes and their local interactions with the medium are sufficiently similar, then the difference in potential between the two electrodes equals the difference in potential between the two points in space. Dividing by the magnitude of the vector distance **d** between them yields the component of the electric field linking the two sensors.

The most symmetrical element is a sphere, and many *in situ* electric field measurements use spherical electrodes mounted on insulating booms that are made as long as financial and mechanical constraints allow. The length is maximized as the voltage signal V_s is proportional to $V_s = -\mathbf{E}' \cdot \mathbf{d}$, whereas the sources of error are either independent of the separation from the spacecraft or decrease drastically with separation distance. Typical boom lengths on sounding rockets range from 1 to 15 m, and on satellites, 100-m tip-to-tip separations are commonly used.

Some of these error sources can be understood from the interaction between a single electrode and the local plasma. Consider first a floating probe, that is, one from which no current is drawn by the attached electronics. Due to their higher velocity, the flux of electrons to a given surface, nv_e, exceeds the flux of ions to the same surface, nv_i. An electrode will thus charge up negatively, repelling just enough electrons that the electron and ion fluxes are equal. For a spherical probe at rest with area A in a plasma of temperature T and with electron and ion masses m and M, the current I to the probe is given by (Fahleson, 1967)

$$I = (Ane/4)(8k_B T/\pi M)^{1/2}$$
$$- (Ane/4)(8k_B T/\pi m)^{1/2} \exp(eV_F/k_B T).$$

Setting $I=0$ for a floating probe and solving this equation for V_F yields

$$V_F = (k_B T/e)\ln(m/M)^{1/2}.$$

For a 1000 K, O^+ plasma, $V_F = -5.1 k_B T/e \approx -0.44$ V. Any asymmetries between the probes will make V_F differ and create an error signal. Sources of errors include asymmetric ion collection due to the spacecraft motion, which is usually at a velocity higher than the ion thermal speed, asymmetric photo-emission, which behaves like an ion current to each probe, and asymmetric electron collection. The latter may arise if the main spacecraft emits photoelectrons that are collected by probes or if the magnetic field lines threading one of the probes pass near the main spacecraft. An additional error signal comes from any possible difference in the average work function of the two surfaces.

A schematic diagram (Mozer, 1973) showing the potentials involved when two separated electrodes with different work functions WF_1 and WF_2, different floating potentials V_1 and V_2, and separated by a distance **d** in an external electric field **E′** are connected to a differential electrometer with input resistance R, as given in Fig. 7.7. To make the measurement, a current I must be drawn by the electronics. This creates an additional potential difference at the two spheres denoted by $R_1 I$ and $R_2 I$, where

$$R_j = (\partial V / \partial I)_{V_F}$$

is the dynamic resistance of the electrode-plasma contact evaluated at the floating potential. Differentiating the voltage equation and evaluating R at the floating potential yields

$$R = (k_B T/e)/I_i.$$

That is, the electron temperature measured in electron volts divided by the ion current to the probe yields the dynamic resistance for a floating probe. For a sphere of 100-cm²-projected collection area, moving through a plasma of temperature 0.16 eV at a velocity of 1 km/s with a density of 10^4 cm^{-3}, $R \approx 10^7 \, \Omega$. Analysis of the potential diagram shows that the electrometer measures

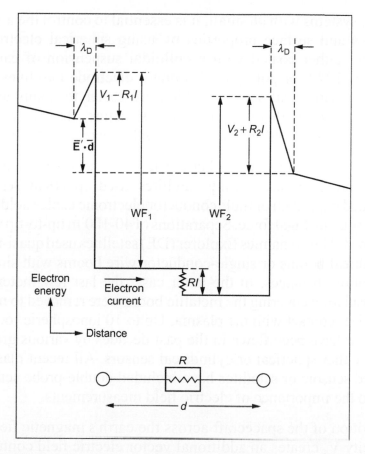

Fig. 7.7 *Electron potential energy as a function of position near a Langmuir double probe. After Kelley (2009). Reproduced with permission of Elsevier.*

$$RI = [\mathbf{E}' \cdot \mathbf{d} + (V_1 - V_2) + (\mathrm{WF}_2 - \mathrm{WF}_1)] / [1 + (R_1/R) + (R_2/R)].$$

(Notice that the potential plotted is the electron potential energy, which is the negative of the normal potential in circuit analysis.) Clearly, to detect the full $\mathbf{E}' \cdot \mathbf{d}$ potential, we need to have $R \gg R_j$ and keep the error terms in the numerator small. This is quite simple with modern electronics where input resistances of $10^{12} \Omega$ are easily available, but was much more problematic in the 1960s.

For systems with $|\mathbf{d}|$ small, it is essential to control the asymmetries and surface properties by using spherical electrodes that are either coated with a colloidal suspension of carbon (Kelley, 1970) or gold plated. Longer electrode baselines can employ cylindrical sensors, which are easier to manufacture and interface with the mechanical structure. These systems overcome the signal-to-noise problem by increasing the signal.

Concerning early ionospheric satellites, the very successful S3-2 and S3-3 and Viking satellites used spherical sensors mounted on triaxial or multiconductor electronic cables held outward by centrifugal force. Separations of 40-100 m tip-to-tip were obtained. The Dynamics Explorer (DE) satellites used quasi-rigid cylindrical booms or single-conductor wire booms with similar separation distances. In the latter case, the last few meters of the insulation covering the metallic boom were removed to make electrical contact with the plasma. Up to 10 ionospheric rockets per year have been flown in the past decades by various groups using either spherical or cylindrical sensors. All recent plasma-related sensors or satellites have included double-probe sensors due to the importance of electric field measurements.

Motion of the spacecraft across the earth's magnetic field at velocity $\mathbf{V_s}$ creates an additional vector electric field contribution as, in the moving frame, the field $\mathbf{E}'$ is given by

$$\mathbf{E}' = \mathbf{E_A} + \mathbf{V_s} \times \mathbf{B}$$

where $\mathbf{E_A}$ is the desired ambient electric field. Subtracting two vectors requires accurate knowledge of $\mathbf{V_s}$ and the vehicle attitude. The magnetic field itself is usually known quite well at ionospheric heights and is not a problem when carrying out the required vector subtraction. This need for accurate attitude information is crucial for measurements below $60°$ latitude.

The balloon-borne double-probe method is essentially identical except that the electronics and the system in general are much more sensitive to the requirement ($R \gg R_j$). This is due to the very low charged-particle density in the atmosphere and the correspondingly high value of R_j. What is surprising is that the horizontal electric field components at balloon heights (30 km) have anything at all to do with the ionosphere. In fact, a number of theoretical studies as well as the data itself show that ionospheric fields of large horizontal scale (±100 km) map nearly unattenuated down to 30 km (Kelley and Mozer, 1975; Mozer and Serlin, 1969). Superpressure balloons have been used in the southern hemisphere where very long duration flights are possible, several months long, in fact. Communication with these balloons is done via satellite links (Holzworth et al., 1984).

7.4.6 Electrostatic Wave Measurements

Almost all of the waves discussed in this text are electrostatic in nature; that is, there are no associated magnetic field fluctuations. From Maxwell's equation,

$$\nabla \times \delta\mathbf{E} = -\partial(\delta B)/\partial t = 0,$$

so $\delta\mathbf{E}$ may be derived from a potential

$$\delta\mathbf{E} = -\nabla(\delta\phi)$$

and hence the term "electrostatic" is used. This term does not imply a static phenomenon, however, as these waves can travel at quite high velocities and display high-frequency fluctuations. The linearized version of this equation is

$$\delta\mathbf{E} = -i\mathbf{k}(\delta\phi),$$

which shows that $\delta\mathbf{E}$ is antiparallel to the propagation vector.

For frequencies below either the ion plasma frequency or the lower hybrid frequency, whichever is lower, both ions and electrons participate in the response of the medium to wave electric fields. This implies further that density fluctuations $\delta n/n$ and a velocity fluctuation δV are associated with the wave. Electrostatic waves may thus be detected by the density probes, by electric field detectors, and by the drift meters described above.

Because of the spacecraft motion relative to the plasma, even pure spatial variations appear as temporal fluctuations in the vehicle frame. If the wave phase velocity in the plasma is much less than the relative velocity, then the frequency ω in the spacecraft frame is just equal to $\mathbf{k} \cdot \mathbf{V_R}$. This assumption is almost always valid in the case of satellite measurements because of the large spacecraft velocities ($\sim$7000 m/s) and is often (but not always) valid for rockets as well. In the auroral zone, plasma drifts can easily exceed the components of the vehicle's rocket velocity and plasma perpendicular to $\mathbf{B}$. In this case, it is important to measure the quasi-dc electric field. In fact, the relative velocity of the vehicle and plasma perpendicular to $\mathbf{B}$ is $\mathbf{E_R} \times \mathbf{B}/B^2$, where $\mathbf{E_R}$ is the electric field measured directly with the rocket instruments in the rocket reference frame. If the wave frequency in the plasma frame is of the order of $\mathbf{k} \cdot \mathbf{V_R}$, the analysis is more complicated. Low-apogee rockets flown to study the equatorial and auroral electrojets fall in this category.

Although probes and radar systems both respond to irregularities in the plasma density, a fundamental difference must be kept in mind when comparing results. As discussed above, the backscatter radar responds to irregularities with a unique wave number $\mathbf{k}$ that is nearly a three-dimensional delta function. A probe responds to any wave number for which $\mathbf{k} \cdot \mathbf{V_R} \neq 0$. Even at a given frequency ω', the wave number is not unique, as there are infinitely many wave numbers, $\mathbf{k}$, such that $\omega' = \mathbf{k} \cdot \mathbf{V_R}$.

If the wavelength of an electrostatic wave is shorter than the separation distance $\mathbf{d}$ between electrodes, then the full electric field potential $-\mathbf{E} \cdot \mathbf{d}$ is not measured. In fact, if an integral number of half-wavelengths separates the sensors, the voltage difference is zero. For a plane wave, the voltage response is

$$V_s = [-\mathbf{E} \cdot \mathbf{d} \cos(\omega t - \theta) \sin(\mathbf{k} \cdot \mathbf{d}/2)]/(\mathbf{k} \cdot \mathbf{d}/2).$$

If $\mathbf{k}\cdot\mathbf{d} \ll 1$, this reduces to $\mathbf{E}\cdot\mathbf{d}$. For $\mathbf{k}\cdot\mathbf{d}/2$ large and a finite spread of wave numbers about some k_0, the sensor response is proportional to the wave potential $\mathbf{E}_0/k_0$ and not the wave electric field. Under certain conditions, these nulls can be used to determine the wavelength of the electrostatic wave as $\mathbf{d}$ is known (Temerin, 1978). Such information is invaluable for sorting out which wave mode causes a given fluctuation.

Simultaneous measurements of $\delta\mathbf{E}(\omega)$ or $\delta\mathbf{V}(\omega)$ and the density fluctuation spectrum $\delta n(\omega)/n$ can sometimes be used to determine the wave mode, as theory can be used to predict the relationship between the various components of the wave (Kelley, 2009).

7.5 BARIUM ION CLOUD MEASUREMENTS

As the electric field is of crucial importance to the understanding of ionospheric dynamics, tracer techniques were developed in the 1960s to measure the flow velocity of the ionospheric plasma. The basic idea is to inject a small number of atoms into the medium under conditions such that the cloud is fully sunlit but observers on the ground are in darkness. For certain materials, the sunlight both ionizes the material and makes it visible via a resonant scattering process. Barium vapor has proven to be the best material for this experiment, and well over 100 releases have been carried out over the years at altitudes from 150 to 60,000 km.

To vaporize the barium metal, very high temperatures are needed (>2000 K), and a thermite reaction is used to attain the required vaporization heat. Small clouds are needed if the tracer aspect is important. "Small" in this case means that the height-integrated conductivity of the cloud must be less than that of the ionosphere. Following Haerendel et al. (1967), at F-region altitudes, the electric field is given by

$$\mathbf{E}_\perp = \frac{1+\lambda^*}{2}B\left[\hat{B} \times \mathbf{V}_\perp + \frac{1}{k_i}\left(\mathbf{V}_\perp - \mathbf{U}_{n\perp}\right) + \frac{\lambda^*-1}{\lambda^*+1}\mathbf{U}_{n\perp} \times \mathbf{B}\right]$$

where $\mathbf{V}_\perp$ is the velocity of the ion cloud perpendicular to $\mathbf{B}$, λ^* is the ratio of height-integrated Pedersen conductivities in the presence and in the absence of the cloud, k_i is the ratio of the gyration frequency to the collision frequencies of a barium ion, $\mathbf{U}_{n\perp}$ is the neutral wind velocity in the reference frame fixed to the earth, $\mathbf{B}$ is a unit vector parallel to the local magnetic field direction, and $\mathbf{E}$ and $\mathbf{E}_\perp$ represent the ionospheric electric field in a frame of reference fixed to the earth in the plane perpendicular to the local magnetic field direction. For $\lambda^* = 1$ and k_i large (F-region altitudes), $\mathbf{V}_\perp$ is equal to $(\mathbf{E} \times \mathbf{B})/B^2$. The method has proven to be extremely valuable for measuring electric fields in the ionosphere. This is particularly true at low latitudes, where probe techniques and drift meters must contend with satellite velocities greatly in excess of the plasma velocity.

Historically, it is interesting to note that, after the nuclear test ban treaty, barium releases were used to study the physics of high-altitude explosions. The reason for this is clear from Fig. 7.8. Here, we show an image of a barium release and nuclear explosion in the ionosphere some minutes after the event. The scales in the figure are identical and the similarity is striking. The km-scale irregularities in both cases are due to the generalized Rayleigh-Taylor instability in both cases. In the barium cloud case, a large cloud is needed (unlike the small cloud case used for electric field measurements) and the destabilizing source is the wind. For the nuclear case, it is the deceleration of the debris cloud due to collisions with the background atmosphere. These irregularities were important due to the fear that, in a nuclear war, they would disrupt communications dramatically. Currently, we are more concerned with natural irregularities (space weather) creating communication and navigation (GPS) system interruptions. The theoretical and experimental work in the nuclear case by the Department of Defense has contributed to our current understanding of the natural case.

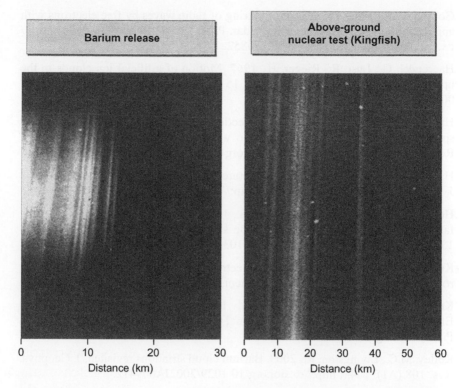

Fig. 7.8 Comparison of a striated barium cloud and a late-time nuclear explosion on the same distance scale. Both events occurred in the 150-200 km height range. After Kelley and Livingston (2003). Reproduced with permission of the American Geophysical Union.

REFERENCES

Bauer, S.J., Nagy, A.F., 1975. Ionospheric direct measurement techniques. Proc. IEEE 63 (2), 230–249. http://dx.doi.org/10.1109/PROC.1975.9733.

Bowles, K.L., 1958. Observations of vertical-incidence scatter from the ionosphere at 41 Mc/sec. Phys. Rev. Lett. 1 (12), 454–455. http://link.aps.org/doi/10.1103/PhysRevLett.1.454.

Chen, F.F., 1984. Introduction to Plasma Physics and Controlled Fusion, second ed. Plasma Physics, vol. 1. Plenum Press, New York.

Fahleson, U.V., 1967. Theory of electric field measurements conducted in the magnetosphere with electric probes. Space Sci. Rev. 7 (2–3), 238–262. http://dx.doi.org/10.1007/BF00215600.

Farley, D.T., 1979. The ionospheric plasma. In: Kennel, C.F., Lanzerotti, L.J., Parker, E.N. (Eds.), Solar System Plasma Physics, vol. 3. North-Holland Publishing Co., Amsterdam, pp. 271–298.

Gordon, W.E., 1958. Incoherent scattering of radio waves by free electrons with application to space exploration by radar. Proc. IRE 46 (11), 1824–1829. http://dx.doi.org/10.1109/JRPROC.1958.286852.

Haerendel, G., Lust, R., Rieger, E., 1967. Motion of artificial ion clouds in the upper atmosphere. Planet. Space Sci. 15 (1), 1–18. http://dx.doi.org/10.1016/0032-0633(67)90062-1.

Hagen, J.B., Behnke, R.A., 1976. Detection of the electron component of the spectrum in incoherent scatter of radio waves by the ionosphere. J. Geophys. Res. 81 (19), 3441–3443. http://dx.doi.org/10.1029/JA081i019p03441.

Hagen, J.B., Hsu, P.Y., 1974. The structure of the protonosphere above Arecibo. J. Geophys. Res. 79 (28), 4269–4275. http://dx.doi.org/10.1029/JA079i028p04269.

Holzworth, R.H., Onsager, T., Powell, S., 1984. Planetary-scale variability of the fair-weather vertical electric field in the stratosphere. Phys. Rev. Lett. 53 (14), 1398–1401. http://link.aps.org/doi/10.1103/PhysRevLett.53.139.

Kelley, M.C., 1970. Auroral zone electric field measurements on sounding rockets. Ph.D. Thesis. Physics Department, University of California at Berkeley.

Kelley, M.C., 2009. The Earth's Ionosphere: Plasma Physics and Electrodynamics, second ed. International Geophysics Series, vol. 96. Academic Press, Burlington, MA.

Kelley, M.C., Livingston, R., 2003. Barium cloud striations revisited. J. Geophys. Res. 108 (A1), 1044. http://dx.doi.org/10.1029/2002JA009412.

Kelley, M.C., Mozer, F.S., 1975. Simultaneous measurement of the horizontal components of the earth's electric field in the atmosphere and in the ionosphere. J. Geophys. Res. 80 (22), 3275–3276. http://dx.doi.org/10.1029/JA080i022p03275.

Mott-Smith, H., Langmuir, I., 1926. The theory of collectors in gaseous discharges. Phys. Rev. 28 (4), 727–763. http://link.aps.org/doi/10.1103/PhysRev.28.727.

Mozer, F.S., 1973. Analysis of techniques for measuring DC and AC electric fields in the magnetosphere. Space Sci. Rev. 14 (23), 272–313. http://dx.doi.org/10.1007/BF02432099.

Mozer, F.S., Serlin, R., 1969. Magnetospheric electric field measurements with balloons. J. Geophys. Res. 74 (19), 4739–4754. http://dx.doi.org/10.1029/JA074i019p04739.

Szuszczewicz, E.P., Takacs, P.Z., 1979. Magnetosheath effects on cylindrical Langmuir probes. Phys. Fluids 22 (12), 2424–2429. http://dx.doi.org/10.1063/1.862556.

Temerin, M., 1978. The polarization, frequency, and wavelengths of high-latitude turbulence. J. Geophys. Res. 83 (A6), 2609–2616. http://dx.doi.org/10.1029/JA083iA06p02609.

INDEX

Note: Page numbers followed by f indicate figures.

Printed and bound by CPI Group (UK) Ltd, Croydon, CR0 4YY

03/10/2024

01040418-0003